In situ characterization of thin film growth

Related titles:

Electromigration in thin films and electronic devices
(ISBN 978-1-84569-937-6)
Electromigration is a significant problem affecting the reliability of microelectronic devices such as integrated circuits. Recent research has focused on how electromigration affects the increasing use by the microelectronics industry of lead-free solders and copper interconnects. Part I reviews ways of modelling and testing electromigration. Part II discusses electromigration in copper interconnects, while Part III covers solder.

Thin film growth
(ISBN 978-1-84569-736-5)
Thin film technology is used in many applications such as microelectronics, optics, magnetics, hard and corrosion-resistant coatings and micromechanics. This book provides a review of the theory and techniques for the deposition of thin films. *Thin film growth* will help the reader understand the variables affecting growth kinetics and microstructural evolution during deposition. Part I covers the theory and modelling of thin film growth while Part II describes the techniques and mechanisms of film growth. This section covers examples such as silicon nanostructured thin films, colloidal crystal thin films and graphene thin films. It also contains discussion of pliable substrates and thin films for particular functions.

Advanced piezoelectric materials
(ISBN 978-1-84569-534-7)
Piezoelectric materials produce electric charges on their surfaces as a consequence of applying mechanical stress. They are used in the fabrication of a growing range of devices such as transducers, actuators, pressure sensor devices and increasingly as a way of producing energy. This book provides a comprehensive review of advanced piezoelectric materials, their properties, methods of manufacture and applications. It covers lead zirconate titanate (PZT) piezo-ceramics, relaxor ferroelectric ceramics, lead-free piezo-ceramics, quartz-based piezoelectric materials, the use of lithium niobate and lithium in piezoelectrics, single crystal piezoelectric materials, electroactive polymers (EAP) and piezoelectric composite materials.

Details of these and other Woodhead Publishing materials books can be obtained by:

- visiting our web site at www.woodheadpublishing.com
- contacting Customer Services (e-mail: sales@woodheadpublishing.com; fax: +44 (0) 1223 832819; tel.: +44 (0) 1223 499140 ext. 130; address: Woodhead Publishing Limited, 80 High Street, Sawston, Cambridge CB22 3HJ, UK)
- contacting our US office (email: usmarketing@woodheadpublishing.com; tel.: (215) 928 9112; address: Woodhead Publishing, 1518 Walnut Street, Suite 1100, Philadelphia, PA 19102-3406, USA)

If you would like to receive information on forthcoming titles, please send your address details to: Francis Dodds (address, tel. and fax as above; e-mail: francis.dodds@woodheadpublishing.com). Please confirm which subject areas you are interested in.

In situ characterization of thin film growth

Edited by
Gertjan Koster and Guus Rijnders

WOODHEAD
PUBLISHING

Oxford   Cambridge   Philadelphia   New Delhi

© Woodhead Publishing Limited, 2011

Published by Woodhead Publishing Limited,
80 High Street, Sawston, Cambridge CB22 3HJ, UK
www.woodheadpublishing.com

Woodhead Publishing, 1518 Walnut Street, Suite 1100, Philadelphia, PA 19102-3406, USA

Woodhead Publishing India Private Limited, G-2, Vardaan House, 7/28 Ansari Road, Daryaganj, New Delhi – 110002, India
www.woodheadpublishingindia.com

First published 2011, Woodhead Publishing Limited
© Woodhead Publishing Limited, 2011
The authors have asserted their moral rights.

This book contains information obtained from authentic and highly regarded sources. Reprinted material is quoted with permission, and sources are indicated. Reasonable efforts have been made to publish reliable data and information, but the authors and the publisher cannot assume responsibility for the validity of all materials. Neither the authors nor the publisher, nor anyone else associated with this publication, shall be liable for any loss, damage or liability directly or indirectly caused or alleged to be caused by this book.

Neither this book nor any part may be reproduced or transmitted in any form or by any means, electronic or mechanical, including photocopying, microfilming and recording, or by any information storage or retrieval system, without permission in writing from Woodhead Publishing Limited.

The consent of Woodhead Publishing Limited does not extend to copying for general distribution, for promotion, for creating new works, or for resale. Specific permission must be obtained in writing from Woodhead Publishing Limited for such copying.

Trademark notice: Product or corporate names may be trademarks or registered trademarks, and are used only for identification and explanation, without intent to infringe.

British Library Cataloguing in Publication Data
A catalogue record for this book is available from the British Library.

Library of Congress Control Number: 2011935503

ISBN 978-1-84569-934-5 (print)
ISBN 978-0-85709-495-7 (online)

The publisher's policy is to use permanent paper from mills that operate a sustainable forestry policy, and which has been manufactured from pulp which is processed using acid-free and elemental chlorine-free practices. Furthermore, the publisher ensures that the text paper and cover board used have met acceptable environmental accreditation standards.

Typeset by Replika Press Pvt Ltd, India
Printed by Lightning Source

Contents

Contributor contact details ix

Part I Electron diffraction techniques for studying thin film growth *in situ* 1

1 Reflection high-energy electron diffraction (RHEED) for *in situ* characterization of thin film growth 3
G. KOSTER, University of Twente, the Netherlands

1.1	Reflection high-energy electron diffraction (RHEED) and pulsed laser deposition (PLD)	3
1.2	Basic principles of RHEED	5
1.3	Analysis of typical RHEED patterns: the influence of surface disorder	8
1.4	Crystal growth: kinetics vs thermodynamics	12
1.5	Variations of the specular intensity during deposition	13
1.6	Kinetical growth modes and the intensity response in RHEED	18
1.7	RHEED intensity variations and Monte Carlo simulations	23
1.8	Conclusions	25
1.9	Acknowledgements	26
1.10	References	26

2 Inelastic scattering techniques for *in situ* characterization of thin film growth: backscatter Kikuchi diffraction 29
N. J. C. INGLE, University of British Columbia, Canada

2.1	Introduction	29
2.2	Kikuchi patterns	30
2.3	Kikuchi lines in reflection high-energy electron diffraction (RHEED) images	33
2.4	Dual-screen RHEED and Kikuchi pattern collection	37

2.5	Lattice parameter determination	40
2.6	Epitaxial film strain determination	41
2.7	Kinematic and dynamic scattering	42
2.8	Epitaxial film structure determination	45
2.9	Conclusion	48
2.10	References	49

Part II Photoemission techniques for studying thin film growth *in situ* — 53

3 Ultraviolet photoemission spectroscopy (UPS) for *in situ* characterization of thin film growth — 55
K. M. SHEN, Cornell University, USA

3.1	Introduction	55
3.2	Principles of ultraviolet photoemission spectroscopy (UPS)	56
3.3	Applications of UPS to thin film systems	63
3.4	Future trends	72
3.5	References	73

4 X-ray photoelectron spectroscopy (XPS) for *in situ* characterization of thin film growth — 75
H. BLUHM, Lawrence Berkeley National Laboratory, USA

4.1	Introduction	75
4.2	*In situ* monitoring of thin film growth	83
4.3	Measuring the reaction of thin films with gases using ambient pressure X-ray photoelectron spectroscopy (XPS)	88
4.4	*In situ* measurements of buried interfaces using high kinetic energy XPS (HAXPES)	92
4.5	Conclusions	94
4.6	Acknowledgments	95
4.7	References	96

5 *In situ* spectroscopic ellipsometry (SE) for characterization of thin film growth — 99
J. N. HILFIKER, J.A. Woollam Co., Inc., USA

5.1	Introduction	99
5.2	Principles of ellipsometry	100
5.3	*In situ* spectroscopic ellipsometry (SE) characterization	109
5.4	*In situ* considerations	119
5.5	Further *in situ* SE examples	132
5.6	Conclusions	143

5.7	Sources of further information and advice	144
5.8	Acknowledgments	146
5.9	References	146

Part III Alternative *in situ* characterization techniques — 153

6 *In situ* ion beam surface characterization of thin multicomponent films — 155
L. V. GONCHAROVA, The University of Western Ontario, Canada

6.1	Introduction	155
6.2	Background to ion backscattering spectrometry and time-of-flight (TOF) ion scattering and recoil methods	157
6.3	Experimental set-ups	161
6.4	Studies of film growth processes relevant to multicomponent oxides	165
6.5	Conclusions	176
6.6	Acknowledgments	176
6.7	References	176

7 Spectroscopies combined with reflection high-energy electron diffraction (RHEED) for real-time *in situ* surface monitoring of thin film growth — 180
P. G. STAIB, Staib Instruments, Inc., USA

7.1	Introduction	180
7.2	Overview of processes and excitations by primary electrons in the surface	181
7.3	Recombination and emission processes	188
7.4	Descriptions and results of *in situ* spectroscopies combined with reflection high-energy electron diffraction (RHEED)	191
7.5	Conclusion and future trends	206
7.6	Sources of further information and advice	209
7.7	References	209

8 *In situ* deposition vapor monitoring — 212
V. MATIAS, Los Alamos National Laboratory, USA and R. H. HAMMOND, Stanford University, USA

8.1	Introduction	212
8.2	Overview of vapor flux monitoring	212
8.3	Quartz crystal microbalance (QCM)	214
8.4	Vapor ionization techniques	217
8.5	Optical absorption spectroscopy techniques	223

8.6	Summary of techniques and resources	230
8.7	Case studies	231
8.8	Conclusions	235
8.9	Acknowledgments	236
8.10	References	236
9	Real-time studies of epitaxial film growth using surface X-ray diffraction (SXRD)	239
	G. Eres, J. Z. Tischler, C. M. Rouleau, B. C. Larson and H. M. Christen, Oak Ridge National Laboratory, USA and P. Zschack, Argonne National Laboratory, USA	
9.1	Introduction	239
9.2	Growth kinetics studies of pulsed laser deposition (PLD) using surface X-ray diffraction (SXRD)	245
9.3	Real-time SXRD in $SrTiO_3$ PLD: an experimental case study	250
9.4	Future trends	267
9.5	Acknowledgment	269
9.6	References	269
	Index	*274*

Contributor contact details

(* = main contact)

Editors

Gertjan Koster
MESA+ Institute for
   Nanotechnology
University of Twente
Carre 3247
PO Box 217
NL-7500AE Enschede
The Netherlands

E-mail: g.koster@utwente.nl

Guus Rijnders
University of Twente
Faculty of Science & Technology
Carre, 3243
PO Box 217
NL-7500 AE Enschede
The Netherlands

E-mail: a.j.h.m.rijnders@utwente.nl

Chapter 1

Gertjan Koster
MESA+ Institute for
   Nanotechnology
University of Twente
Carre 3247
PO Box 217
NL-7500AE Enschede
The Netherlands

E-mail: g.koster@utwente.nl

Chapter 2

N. J. C. Ingle
Advanced Materials and Process
   Engineering Laboratory
University of British Columbia
2355 East Mall
Vancouver
BC V6T 1Z4
Canada

E-mail: ingle@physics.ubc.ca

Chapter 3

Kyle M. Shen
Laboratory of Atomic and Solid
  State Physics
532A Clark Hall
Department of Physics
Cornell University
Ithaca
NY 14853
USA

E-mail: kmshen@cornell.edu

Chapter 4

H. Bluhm
Chemical Sciences Division
Lawrence Berkeley National
  Laboratory
Mail Stop 6R2100
Berkeley
CA 94720
USA

E-mail: hbluhm@lbl.gov

Chapter 5

J. N. Hilfiker
J.A. Woollam Co., Inc.
645 M Street
Suite 102
Lincoln
NE 68508
USA

E-mail: jhilfiker@jawoollam.com

Chapter 6

L. V. Goncharova
Department of Physics and
  Astronomy
The University of Western Ontario
1151 Richmond Street
London
Ontario
N6A 3K7
Canada

E-mail: lgonchar@uwo.ca

Chapter 7

P. G. Staib
Staib Instruments, Inc.
101 Stafford Court
Williamsburg
VA 23185
USA

E-mail: pstaib@staibinstruments.com

Chapter 8

V. Matias
Los Alamos National Laboratory
Mail Stop T004
Los Alamos
NM 87545
USA

E-mail: vlado@lanl.gov

R. H. Hammond*
Stanford University
Stanford
CA 94305
USA

E-mail: rhammond@stanford.edu

Chapter 9

G. Eres*, J. Z. Tischler, C. M. Rouleau, B. C. Larson and H. M. Christen
Materials Science and Technology Division
Oak Ridge National Laboratory
Oak Ridge
TN 37831
USA

E-mail: eresg@ornl.gov

P. Zschack
X-Ray Science Division
Advanced Photon Source
Argonne National Laboratory
USA

Part I
Electron diffraction techniques for studying thin film growth *in situ*

1
Reflection high-energy electron diffraction (RHEED) for *in situ* characterization of thin film growth

G. KOSTER, University of Twente, the Netherlands

Abstract: In this chapter reflection high-energy electron diffraction (RHEED) is described in combination with pulsed laser deposition (PLD). Both the use of RHEED as a real-time rate-monitoring technique as well as methods to study the nucleation and growth during PLD are discussed. After a brief introduction of RHEED and demonstration of typical surface diffraction patterns, a case will be made for the step-density model to describe the intensity variations encountered during deposition. Finally, an overview of these intensity variations, the intensity response during a RHEED experiment as a result of various kinetic growth modes, will be given.

Key words: thin films, pulsed laser deposition, surface electron diffraction, reflection high-energy electron diffraction (RHEED).

1.1 Reflection high-energy electron diffraction (RHEED) and pulsed laser deposition (PLD)

Reflection high-energy electron diffraction (RHEED) was limited to low background pressures only until the development of high-pressure RHEED, which made it possible to monitor the surface structure *in situ* during the deposition of oxides at higher pressures, presented new possibilities (Rijnders *et al.*, 1997). Figure 1.1 is a schematic picture of a typical high-pressure RHEED set-up. Besides observed intensity oscillations due to layer-by-layer growth, enabling accurate growth rate control, it has become clear that intensity relaxation observed due to the typical pulsed way of deposition leads to a wealth of information about growth parameters (Blank *et al.*, 1998).

Pulsed laser deposition (PLD) has become an important technique in the fabrication of novel materials. Starting in the mid-1960s (Ready, 1963), when the first attempts to produce high-quality thin films showed the promise of this technique, it has taken the discovery of high-T_c superconductors for PLD to become widespread. The main advantages of PLD are the relatively easy stoichiometric transfer of material from the target to the substrate and an almost free choice of (relatively high) background pressure. For instance,

4   *In situ* characterization of thin film growth

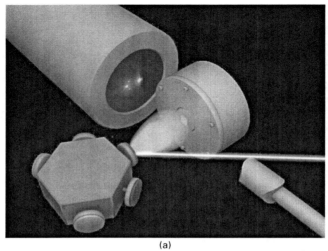

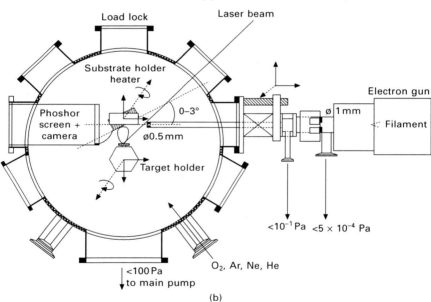

1.1 (a) and (b) A schematic view of a pulsed laser deposition chamber equipped with high-pressure RHEED.

during the deposition of oxides, an oxygen background pressure up to 1 mbar is usually used.

The articles by Christen and Eres (2008) and Rijnders and Blank (2007a) give an extensive overview of the various growth models that apply to the highly kinetic type of deposition.

1.2 Basic principles of RHEED

In a typical RHEED system, a high-energy electron beam (10–50 keV) arrives at a surface under a grazing incident angle (0.1–5°); see also Fig. 1.2a. At these energies the electrons can penetrate any material for several hundreds of nanometres. However, due to the grazing angle of incidence, the electrons only interact with the topmost layer of atoms (1–2 nm) at the surface, which makes the technique very surface sensitive. In contrast, low-energy electron diffraction (LEED), is surface sensitive due to the low penetration depth of low energy electrons (100–500 eV). The scattered electrons collected on a phosphorus screen form a diffraction pattern characteristic for the crystal structure of the surface, and also contain information concerning the morphology of the surface. As we will see, RHEED is remarkably sensitive to variations in morphology and roughness during thin film growth and therefore is often used as method to monitor the real-time thickness.

Examples of other surface-sensitive techniques, which are used to probe crystal structure and morphology, are X-ray diffraction (see, for example, for a discussion of disadvantages and advantages of different diffraction techniques; Yang, 1993; Christen and Eres, 2008) and scanning probe microscopy (SPM); see Chapter 2. The choice for RHEED is based on both the accessibility of

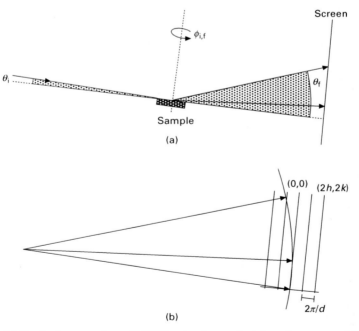

1.2 Typical set-up for a RHEED experiment in (a) real space and (b) reciprocal space.

the sample during a deposition experiment (particle plume perpendicular to the surface and electron beam from the side) and the possibility of using it in a relatively high-pressure environment, as mentioned before. The main reason why RHEED is popular in combination with high vacuum deposition techniques is the observation of intensity oscillations of the specular intensity during deposition (Neave *et al.*, 1983). Intuitively, intensity oscillations can be understood by considering a layer-by-layer growth front. Material deposited on an initially flat surface leads to roughening and a decrease in intensity, whereas upon completion of a crystal layer the surface becomes smoother again, accompanied by a rise in intensity, i.e. periodic island nucleation, growth and coalescence. Accordingly, RHEED can be used as a thickness monitor, since ideally the oscillation period corresponds to the deposition of one crystal layer.

1.2.1 Scattering of electrons by a solid

Electrons interact with the electrostatic potential $V(\vec{r})$ of a solid. To describe the scattering processes, in particular the scattering of a grazing incidence electron beam as in RHEED, one has to distinguish between electrons that lose energy after scattering (inelastic scattering) and electrons that retain their energy (elastic scattering). Examples of features in a RHEED pattern which are caused by inelastically scattered electrons are Kikuchi patterns originating from electrons that have lost some energy while channelling through the crystal; examples are indicated with arrows in Fig. 1.3 (a), (b). The majority of elastic electrons scatter at the outermost atomic layer (Arrot, 1994), i.e. they are the most surface sensitive and the most interesting for a surface scientist. However, inelastic electrons are often indistinguishable from the elastic ones and contribute significantly to the total detected intensity. Many calculations are based only on elastically scattered electrons. If one then assumes weak scattering, i.e. that the scattering cross-section of individual scatterers (atoms) is low (the higher the energy of the incoming electrons, the lower the cross-section (Beeby, 1989), kinematic theory or the first Born approximation is applicable. The incident and scattered electrons are treated as plane waves,

$$I_0 \exp(i\vec{k}_0 \cdot \vec{r}) \qquad [1.1]$$

with $\vec{k}_0 = (\vec{k}_\parallel, \vec{k}_\perp)$ the wave vector. The amplitude of the scattered wave for a certain direction $\vec{k}_0 = (|\vec{k}_0| = |k_s| = 2\pi/\lambda$ where λ is the wavelength of the electrons) is then given by integrating the Fourier components of $V(\vec{r})$ over the volume of the sample,

$$F(\vec{k}) = \int_{-\infty}^{\infty} V(\vec{r}) e^{i(\vec{k} \cdot \vec{r})} dr \qquad [1.2]$$

Reflection high-energy electron diffraction (RHEED)

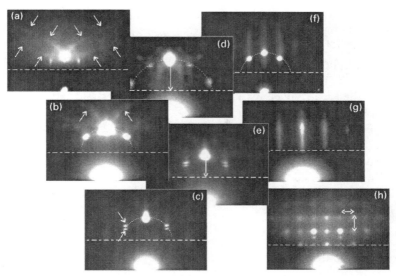

1.3 Typical RHEED patterns for different surface morphologies: (a) as received, (b) treated (with the beam parallel to the terraces), (c) treated (with the beam perpendicular to the terrace), treated (beam perpendicular to the terraces with (d) in-phase and (e) out-of-phase incoming beam, SrTiO$_3$ (001). RHEED patterns of a SrTiO$_3$ surface after deposition of (f) ~one unit-cell layer SrO, (g) several unit-cell layers SrCuO$_2$ and (h) >2 unit-cell layers CaCuO$_2$.

the intensity is proportional to

$$F^*(\vec{k})F(\vec{k}) \qquad [1.3]$$

where $\vec{k} = \vec{k}_s - \vec{k}_0$ is the scattering vector or momentum transfer vector. In the first Born approximation it is possible to replace the potential by the electron density (Arrot, 1994). The problem to be solved is then analogous to X-ray diffraction; the electron density can be thought of as being concentrated in points located at the atomic positions in a solid which act as point scatterers. The scattering power of each atom is given by the atomic scattering factor or form factor; see, for example, Arrot (1994) and Beeby (1989). It is independent of momentum transfer in case of a point scatterer.

Summation over all the atoms in the sample multiplied by the form factor gives the amplitude of the scattered wave and turns out to be non-zero at distinct values of $\vec{k}$, i.e. the reciprocal lattice, when the sample is an infinitely large crystal built from regularly spaced atoms. Usually the Ewald sphere construction, a sphere with radius $2\pi/\lambda$ (elastic scattering) centred on the tip of the wave vector in the reciprocal space of the incoming beam, is used to identify the diffraction angles for a specific geometry and reciprocal lattice. In Fig. 1.2(b) an example of this construction is shown for a singular (i.e.

perfect two-dimensional (2D) lattice of point scatterers) surface in case of RHEED. The reciprocal lattice of such a surface is a set of parallel, regularly spaced, infinitely thin rods perpendicular to the surface which intersect the Ewald sphere at points lying on concentric circles, called the Laue circles. In Fig. 1.3(b) the 0th-order Laue circle, i.e. intersections of the ($0k$) rods with the Ewald sphere, is visible for a $SrTiO_3$ substrate surface, which is an example of a perfect surface.

Following the kinematical approach, surface disorder can be treated as a deviation from a singular surface in either a lateral or a vertical direction resulting in deviations from the reciprocal lattice rods of a perfect surface. Yang (1993) and Lagally *et al.* (1989) give a complete overview.

It turns out that the above is sufficient for a qualitative description of the observed phenomena, e.g. roughening during thin film growth. For a quantitative treatment, necessary for crystallographic structure determination for example, one has to revert to a dynamic theory; first of all, because kinematic theory is in contradiction with the fact that elastically scattered electrons mainly interact with the outermost atoms, whereas weak interaction is assumed, and secondly, due to the grazing incidence, many atoms are involved in the scattering process, suggesting a high probability for multiple-scattering effects (Kawamura, 1989).

Many dynamic calculations aim for the solution of the Schrödinger equation in the one-electron approximation and the right choice of scattering potential $V(\vec{r})$. An imaginary part of $V(\vec{r})$ is added to account for some of the inelastic scattering effects. Since a quantitative treatment is beyond the scope of this chapter please refer to Beeby (1989) for further discussion.

1.3 Analysis of typical RHEED patterns: the influence of surface disorder

1.3.1 Geometrical information of a RHEED pattern

From a RHEED pattern of a perfect low index plane with the beam directed along a low index direction one can determine the in-plane lattice constants as follows: with θ_i and θ_f the angles with respect to the surface, ϕ_i (=0) and ϕ_f the azimuths (with an in-plane principal crystal direction as reference), of the incoming and scattered beam, respectively (see Fig. 1.2a),

$$k_{\parallel \text{beam}} = \frac{2\pi}{\lambda}(\cos\theta_f \cos\phi_f - \cos\theta_i \cos\phi_i) \qquad [1.4a]$$

$$k_{\perp \text{beam}} = \frac{2\pi}{\lambda}(\cos\theta_f \sin\phi_f - \cos\theta_i \sin\phi_i) \qquad [1.4b]$$

$$k_z = \frac{2\pi}{\lambda}(\sin\theta_f + \sin\theta_i) \approx \frac{2\pi}{\lambda}(\theta_f + \theta_i) \text{ (for small angles)} \qquad [1.5]$$

For reflections fulfilling the Bragg condition the lattice constants can be derived using (1.4), for small angles:

$$\frac{n}{d_x} = \frac{1}{\lambda}(\cos\theta_f - \cos\theta_i) \qquad [1.6]$$

$$\frac{n}{d_y} = \frac{1}{\lambda}(\cos\theta_f \sin\phi_f) \qquad [1.7]$$

where d_x and d_y are the in-plane lattice constants seen parallel and perpendicular to the beam, respectively, and n is the order of the reflection. The angles can be determined directly by dividing the relative on-screen distances by the sample-to-screen distance R_s, assuming only small angles. The wavelength λ of the electrons with an energy E (eV), used in eqs [1.4] to [1.7] is approximately given by:

$$\lambda(\text{Å}) = \sqrt{\frac{150}{E}} \qquad [1.8]$$

For the reasons given in section 1.1, the sample-to-screen distance R_s is kept as low as possible. This has the disadvantage that determination of reflection angles is rather inaccurate (relative error ~10%, 5% uncertainty in the sample-to-screen distance and 5% uncertainty for on-screen distances). To eliminate the error in the sample-to-screen distance, one can calibrate the measurements to a known in-plane lattice parameter, for example, bare $SrTiO_3$ (~3.905 Å, ~3.93 Å at 800 °C).

To understand the influence of surface disorder on the RHEED intensity, the problem of scattering from a two-level system, i.e. a certain distribution of islands of height d on a flat surface, will be discussed first. The diffracted intensity is given by:

$$I(\vec{k}) = \int C(\vec{r}) e^{i\vec{k}\cdot\vec{r}} d^3r \qquad [1.9]$$

where $C(\vec{r})$ is the pair correlation function, i.e. the probability of finding two scatterers at a distance r. More explicitly for a two-level system with coverage θ:

$$I(\vec{k}) = [\theta^2 + (1-\theta)^2 + 2\theta(1-\theta)\cos(k_z d)]2\pi\delta(k_x)$$
$$+ 2C'_{ii}(k_z)[1 - \cos(k_z d)] \qquad [1.10]$$

where k_x is given by (1.4) and k_z by (1.5). The intensity consists of a delta peak, corresponding to the Bragg condition superimposed on a broadened diffuse peak, corresponding to disorder (C'_{ii} is the Fourier transform of the reduced correlation function between atoms on the same level (Lent and Cohen, 1984) and depends on the step distribution function). In practice, the delta peak in eq. [1.10] has to be convoluted by the instrument's response function (multiplication with the transfer function (Jorritsma, 1997) in real space) to take instrumental broadening into account. In association with the transfer function a transfer width can be defined, analogous to the coherence length discussed below.

According to this model, intensity is most sensitive for disorder on the surface for the out-of-phase condition,

$$k_z d = (2n + 1)\pi \qquad [1.11]$$

the central spike becomes zero at half coverage.

1.3.2 Examples

In Fig. 1.3, several RHEED patterns are shown of typical surfaces that have been encountered in our experiments. The different features are discussed here.

Prior to any diffraction experiment, the sample is heated to 700°C in vacuum ($\sim 10^{-7}$ mbar) to remove any contamination from the surface. If this procedure is not applied, usually charging effects due to the insulating character of the $SrTiO_3$ substrate destroy the diffraction pattern. In Fig. 1.3(a) a pattern is shown before any heat treatment in vacuum. Although a clear diffraction pattern is visible, the background intensity is rather high and the (0 ± 1) spots in particular are much weaker compared with Fig. 1.3(b) due to surface contamination.

In Fig. 1.3(c) a RHEED pattern is shown of a vicinal $SrTiO_3$ (001) surface, a special case of a 'disordered' surface, where the terraces lie almost parallel to one of the principal axes of the crystal. The incoming beam is directed perpendicular to the terraces and the incidence angle is 1.6°. For 20 keV electrons used here, this is the out-of-phase condition [1.11] for $SrTiO_3$ (001) with only steps of one unit-cell. For a vicinal surface, with the beam perpendicular to the terraces directed down the staircase, the miscut angle is given by (Pukite et al., 1989):

$$1/\theta_c = \frac{\lambda}{d\Delta\theta_f \langle\theta_f\rangle} - \frac{1}{\langle\theta_f\rangle} \qquad [1.12]$$

where $\Delta\theta_f$ is the splitting of the Bragg peak due to the additional periodicity introduced by the existence of terraces, indicated with arrows for the (01) peak in Fig. 1.3. Here we determined a miscut of ~0.25°. Actually, Fig. 1.3(b) is of the same surface, only with the beam directed parallel to the terraces; no splitting is observed. When eq. [1.11] is not satisfied, e.g. by changing the incidence angle, no splitting is observed. Figures 1.3(d) and (e) are patterns of a vicinal $SrTiO_3$ surface at an in-phase and out-of-phase condition, respectively.

For a certain miscut the terraces become 'invisible' to the diffractometer due to instrumental broadening. More generally a surface with disorder over a surface area larger than the coherence area (Van der Wagt, 1994) appears to be flat. The critical terrace length parallel to the incoming electron beam for our diffractometer is estimated to be 3000 Å (miscut angle <0.07°); the coherence length along the beam is an integer multiple of this length (since there must be more than one terrace for interference to occur). In contrast, the coherence length perpendicular to the electron beam is much smaller, i.e. substrates with terraces of <800 Å (miscut angle >0.3°) still appear to be flat in the parallel beam configuration (no splitting $\Delta\phi_f$ has been observed on our substrates). The difference in coherence length parallel and perpendicular to the beam can be explained by the grazing incidence angle geometry typical for RHEED (see, for example, Lagally *et al.*, 1989; Pukite *et al.*, 1989; Van der Wagt, 1994).

In RHEED, any disorder will cause broadening of spots, which will be stronger in the direction of the beam, for similar reasons just mentioned. These so-called streaky patterns are often mistaken for a RHEED pattern of a perfect surface (Lagally *et al.*, 1989).

Figures 1.3(f),(g) and (h) show surfaces with increasing roughness, after deposition of a unit-cell layer of SrO, after deposition of several layers of $SrCuO_2$ and after deposition of several layers of $CaCuO_2$, respectively. Although the SrO surface is still very smooth, some streaking has occurred due to some inevitable disorder, whereas the $SrCuO_2$ surface is disordered and gives a streakier pattern. Again, owing to the grazing-incidence geometry, every detail on the surface responsible for broadening of the reciprocal lattice rods appears to be inflated along the beam direction.

The roughness of the $CaCuO_2$ surface is such that transmission takes place through small asperities on the surface. The pattern has changed from a spotty pattern, where spots lie on circles, to a spotty pattern where the spots form a rectangular pattern. From the positions we conclude that these asperities still have the expected tetragonal $CaCuO_2$ phase.

In conclusion, although information on the atomic structure is very difficult to extract from a RHEED pattern because the intensities cannot simply be deduced using kinematic considerations, the state of the surface can be qualitatively inferred. Particularly during the deposition, RHEED is a powerful tool to monitor the state of the surface and, for example, allows one to interrupt growth when undesired phases of outgrowth are being formed. In addition, monitoring the specular intensity variations, e.g. the intensity oscillations, during a deposition makes it possibile to judge the growth mode of the material, which is discussed below. First a brief explanation is given about the growth models that apply.

1.4 Crystal growth: kinetics vs thermodynamics

Traditionally, thin film growth and crystal growth are described by thermodynamic growth modes, layer-by-layer, layer-by-layer followed by three-dimensional (3D) growth, and simply 3D growth (Volmer–Weber, Stransky–Krastanov and Frank–van der Merwe, respectively), assuming (near) thermodynamic equilibrium. For many deposition techniques, especially in the case of PLD, the latter is not valid and the kinetics of the arriving species has to be taken into account.

In the case of homoepitaxy, kinetic factors determine the growth mode, whereas in the case of heteroepitaxy also thermodynamic factors, e.g. misfit, are also important. In fact, layer-by-layer growth is always predicted for homoepitaxy, even from a thermodynamic point of view (Rosenfeld et al., 1997). However, during deposition of different kinds of materials, i.e. metals, semiconductors and insulators, independently of the deposition technique, a roughening of the surface is observed. Assuming only 2D nucleation, determined by the supersaturation (Markov, 1995), limited interlayer mass transport results in nucleation on top of 2D islands before completion of a unit-cell layer. Still, one can speak of a 2D growth mode. However, nucleation and incorporation of adatoms at step edges proceeds on an increasing number of unit-cell levels, which results in the RHEED intensity oscillations being damped. In fact an exponential decay of the amplitude is predicted assuming conventional molecular beam epitaxy (MBE) deposition conditions (Yang et al., 1995).

To understand the implications of the characteristics of PLD on growth, which are expected to be kinetic in origin, homoepitaxy is the perfect system to study. However, even in the case of heteroepitaxy, kinetic models seem quite appropriate: after some 'transient' behaviour during the first few deposited monolayers (see also below), quasi-homoepitaxial growth is observed.

In order to be able to create a crystal structure by depositing consecutive unit-cell layers of different materials, a layer-by-layer growth model is a prerequisite: nucleation of each next layer may only occur after the previous layer is completed. Note that a 2D growth mode can either be layer-by-layer or step-flow. However, in case of step-flow growth, rate control is not possible. Occasionally, the deposition conditions such as the substrate temperature and ambient gas pressure (oxygen in the case of oxide materials) can be optimized for true 2D growth, e.g. this is the case for homoepitaxy on $SrTiO_3$ (001).

In situ RHEED studies of kinetics of growing systems have been used, as explained by the following examples. First, the transition from step-flow growth to layer-by-layer growth on vicinal surfaces has been used to estimate the diffusion parameters (Neave et al., 1985). With the beam directed parallel to the terraces, intensity oscillations disappear at a critical temperature, when all the material is incorporated at the vicinal steps. The critical diffusion length is then of the order of the terrace width. This simple picture has been extended to include, for instance, non-linear effects (Myers-Beahton and Vvedensky, 1990, 1991) and the fact that the transition is not sharp (Zandvliet *et al.*, 1991; Shitara *et al.*, 1992a, 1992b).

Secondly, another approach was adopted by Vvedensky *et al.* (1990), who studied the relaxation of the RHEED intensity after interruption of the MBE growth. Comparing the characteristic relaxation times observed during experiments with Monte Carlo simulations, the participating microscopic events can be identified, provided that all the relevant mechanisms are included in the simulation model. Note that this approach is closely connected with PLD, since after each laser pulse, relaxation of the RHEED intensity is observed and can be viewed as a kind of interrupted growth.

Finally, the relaxation behaviour of the RHEED intensity after each laser pulse has been studied in case of PLD of $YBa_2Cu_3O_7$ (Karl and Stritzker, 1992), SrO (Nishikawa *et al.*, 1997) and $SrTiO_3$ (Blank, 1999) on $SrTiO_3$. By time-resolved RHEED, the relaxation times for different temperatures have been measured and an estimate of the diffusion barrier is obtained (Rijnders and Blank, 2007b).

1.5 Variations of the specular intensity during deposition

Here we will consider the variations of specular intensity as a function of the variation of the surface roughness during deposition and growth of thin films.

1.5.1 Origin of the specular intensity variations

For homoepitaxy of $SrTiO_3$ the recorded intensity oscillations are depicted in Fig. 1.4(a) together with the calculated intensities using two different models. Similar oscillations have been observed during deposition of GaAs (Van Hove *et al.*, 1983) and Si (Sakamoto *et al.*, 1986). From the shape and amplitude we conclude that at the conditions used here, $SrTiO_3$ deposition proceeds in a true layer-by-layer growth mode, see also Fig. 1.4(d).

For a qualitative understanding one can first turn to the kinematic model (1.10) for a two-level system, exemplified in Fig. 1.4(e) where interference of beams scattered from different levels is the cause of intensity variations. When a crystal layer is filled according to a geometrical distribution of steps, the calculated intensity shows cusp-like oscillations as given in Fig. 1.4(b) (Lagally *et al.*, 1989). The integral intensity depends only on the coverage θ. For $\theta = 0.5$ and at an out-of-phase condition (1.11), a minimum intensity is expected. When only the central delta peak is measured and the

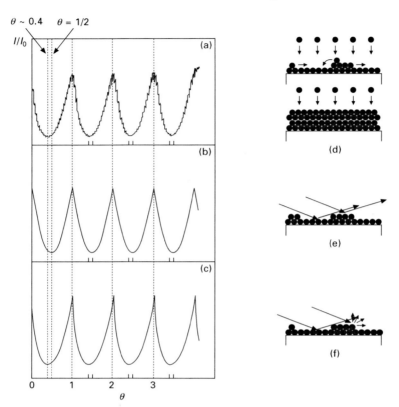

1.4 (a) Intensity oscillations during homoepitaxial growth of SrTiO$_3$ at 850 °C and 3 Pa, indicative of true layer-by-layer growth (d). Calculated intensity oscillations using (b) the diffraction model, of which a schematic representation is given in (e), and (c) the step-density model, of which a schematic representation is given in (f). The number of pulses needed to complete one unit-cell layer is estimated to be 27.

diffuse background is subtracted, this minimum should be zero. When more than two levels are participating in the diffraction process, the calculations become more complicated. However, the basic outcome is that the amplitude of the intensity oscillations decreases with the number of levels involved (Kawamura, 1989); see also below.

Another widely used model is the step density model, where the specular intensity depends negatively on the number of up and down steps on the surface. Every step can act as a diffuse scatterer and the entailed intensity is 'lost' for specular reflection; see Fig. 1.4(f). Although this model is highly empirical and there is no diffraction model to act as a physical explanation, it can qualitatively describe some of our results quite well. Following Stoyanov and Michailov (1988) and assuming that nucleation of islands takes place only on $t = T$ (the period for one unit-cell layer coverage), the step density evolution as a function of the coverage is given by:

$$\sigma(t) = 2\sqrt{\pi N_0}(1 - \theta)\sqrt{-\ln(1 - \theta)} \qquad [1.13]$$

where N_0 is the initial number of nuclei on $t = T$. When growth takes place at only one level and with a constant supply of particles, $\theta \propto t$, and the step density oscillations have again a cusp-like shape; see Fig. 1.4(c). A maximum of the step density σ is expected at $\theta \sim 0.4$ (minimum intensity). For a multi-level system, θ_m are coupled through a series of m rate equations, where m represents the mth unit-cell level. The total step density is given by a summation over all participating levels. The basic outcome is again a decrease of the amplitude of step density oscillations, as the number of involved levels m is increased.

The shape and the occurrence of a minimum for the measured intensity oscillation suggest that here the step-density model is applicable, as indicated in Figs 1.4(a),(b) and (c). The fact that we used an incident angle corresponding to the in-phase condition and still observe strong intensity oscillations also favours the step-density model.

The observation of oscillations depends primarily on the deposition conditions, i.e., the growth mode of the material under investigation (Neave et al.,1985; Terashima et al., 1990; Hirata et al., 1993; Karl, 1993). Also the phase can be determined by the growth conditions (Zandvliet et al., 1991). For a complete overview, see Lagally et al. (1989). Any model describing the intensity oscillations quantitatively should ideally account for diffraction effects and fully describe the microscopic growth mechanisms. In computer simulations using a Monte Carlo algorithm (see Section 1.7) the step-density model is often used to calculate the intensity from a surface model that has been generated by applying microscopic growth models (Vvedensky and Clarke, 1990; Shitara et al., 1992c; Achutharaman et al., 1994; Heyn et al., 1997).

Qualitatively, these simulations seem to describe the growth front for different growth conditions rather well, although the prediction for a damping coefficient by Stoyanov and Michailov is not applicable for strongly damped oscillations. For example, the observations for $SrTiO_3$ deposited at higher pressures or for $YBa_2Cu_3O_7$ (Terashima et al., 1990) resemble more the kinematically calculated intensity of a surface as the outcome of dynamic scaling models for conservative growth conditions according to the Wolf–Villain model (Barabasi, 1995); an exponential decay of the oscillation amplitude is predicted (Yang et al., 1995).

In principle, starting with a perfect surface any damping of oscillations can be ascribed to a transition from a two-level growth front to a multilevel growth front (Lent and Cohen, 1984; Stoyanov and Michailov, 1988). The moment of island nucleation and the moment of coalescence no longer coincide and islands are formed on top of lower level islands, which have not yet coalesced.

Studying intensity oscillations, the main challenge is still the discrimination of diffraction effects from growth-governed effects (Zhang et al., 1987; Lehmpfuhl et al., 1991; Dudarev et al., 1995; Korte and Maksym, 1997; Mitura et al., 1998; Papajova and Vesely, 1998). To minimize dynamical effects, Van der Wagt (1994) suggested monitoring the specular intensity in a slightly misaligned condition.

Another important application of RHEED intensity monitoring is the observation of other growth modes, apart from layer-by-layer modes, and the study of diffusion of material during deposition.

1.5.2 Other growth-induced variations of the specular intensity

In Fig. 1.5, intensity changes typical for pulsed deposition are depicted. As can be seen from this figure, the oscillations are modulated by the laser pulse. The intensity decreases significantly directly after the laser pulse followed by an exponential rise caused by re-crystallization of initially disordered material as reported by, for example, Karl (1993) and Achutharaman et al. (1994). The characteristic times involved contain information about the diffusion and growth on the surface under study. One could suspect that the intensity modulation is just caused by scattering of the electrons by the dense laser plume. However, there are several arguments that rule this out. A large relaxation response is only observed when the surface is relatively smooth, for obvious reasons, e.g., in a 3D pattern as in Fig. 1.3(h), the intensity is not affected by the laser pulse (see also the discussion below). Occasionally, under specific diffraction conditions, the intensity rises after each pulse and finally, the plume 'lifetime' is probably too short to have any effect on the measured intensity. Note that the sampling rate of the CCD camera frame grabber combination is maximally 15 Hz.

Also from the above, it becomes immediately clear that there is an integer number of pulses available for each unit-cell layer (in the event of layer-by-layer growth), which is not necessarily the exact number of pulses needed to complete one unit-cell layer. Sometimes one can observe an aliasing effect caused by the aforementioned discrepancy.

Superimposed on the intensity oscillations due to layer-by-layer growth and the modulation of the laser pulse, a third periodical variation is seen, as indicated in Fig. 1.5(b). The inset shows the calculated intensity according to eq. (1.10), where choosing an appropriate sampling frequency could simulate the aliasing effect.

The relaxation behaviour of the RHEED intensity during homoepitaxy of $SrTiO_3$ can be studied as a function of the deposition temperature and

Reflection high-energy electron diffraction (RHEED)

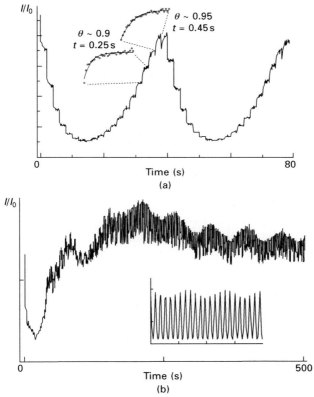

1.5 (a) Modulation of the specular RHEED intensity due to the pulsed way of deposition (T = 750 °C, p_{O_2} = 3 Pa); the insets give enlarged intensity after one laser pulse plus the fit with eq. [1.15] to give characteristic relaxation times. Here, a deposition rate of 19 pulses per unit-cell layer was inferred. (b) Aliasing effect due to the integer number of laser pulses applied, the inset gives the calculated intensity according to a two-level model, from the aliasing period one can deduce ~8.9 pulses per crystal layer.

pressure. In the regime where true layer-by-layer growth or true step-flow growth is observed, a simple model for diffusion of atoms on the surface can be applied, an extension of the known steady-state theory (Mutaftschiev, 1990), which accounts for the measured characteristic relaxation times. From these measurements an activation energy for diffusion could be derived (Rijnders and Blank, 2007a).

In Section 1.5.1, we argued that the oscillations correspond to the fluctuations of the step density on the surface, assuming that nucleation takes place exactly when each unit-cell layer is completed and N_0 nuclei are formed. The step density is then given by eq. [1.16]. In addition, if this model is applicable, only diffusion of material towards ledges needs be considered. Therefore, we assume that, in this growth regime, the relaxation behaviour of the intensity after each laser pulse can be attributed to diffusing particles on the surface, treating each particle as a diffuse scatterer. In the case of diffraction theory, Van der Wagt (1994) derived an expression for the contribution of the adatom density on the surface to the intensity, which is zero for the in-phase condition. The intensity during relaxation is approximately given by:

$$I \propto 1 - \bar{n}_s(t) \qquad [1.14]$$

where the adatom density is averaged over the surface of the circular island. Using this approach we can extract the kinetic factors important for layer-by-layer growth (Koster et al., 1998a). When a step flow-like growth mode is observed, the relaxation is expected to be a function of the adatom density on the terraces, now determined by the average terrace width on the surface.

At a given coverage, r_0 now represents an average maximum distance for the material to diffuse in order to find a ledge (the mean field approximation). In the first approximation, r_0 is considered to be constant for one laser pulse (for this to be true, the amount of deposited material per laser pulse has to be sufficiently small). We evaluated the relaxation times by fitting the intensity for each laser pulse for different coverages, temperatures and pressures using eq. [1.14]. Provided that the assumptions for layer-by-layer growth are still valid, we can use a function of the form:

$$I \sim I_0 \left[1 - \exp\left(-\frac{t}{\tau}\right)\right] \qquad [1.15]$$

where τ gives the characteristic relaxation time. For some examples of fits with eq. [1.15] see the insets in Fig. 1.5(a). For a more recent review of kinetical growth modes during PLD and important parameters, the texts by Christen and Eres (2008) and Rijnders and Blank (2007a) are recommended.

1.6 Kinetical growth modes and the intensity response in RHEED

In this section we will give an overview of the most common kinetical growth modes that one encounters during PLD: step-flow, step-flow-like (steady state), layer-by-layer, multi-level and island growth. In Fig. 1.6 the RHEED intensity response is given, together with a schematic of the growth mode: the RHEED pattern typically seen after a steady state has been reached and the accompanying surface morphology as imaged by atomic force microscopy (AFM). The horizontal axis in this image represents the ratio of the average step density and the average step density during one laser pulse.

First, in Fig. 1.6(a) a (nearly) true step-flow growth mode is shown. All deposited adatoms diffuse to and are captured by existing unit cell steps on

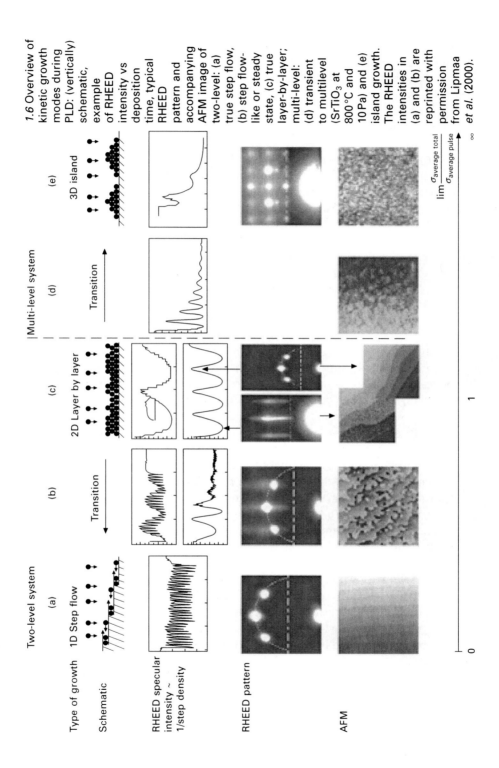

1.6 Overview of kinetic growth modes during PLD: (vertically) schematic, example of RHEED intensity vs deposition time, typical RHEED pattern and accompanying AFM image of two-level: (a) true step flow, (b) step flow-like or steady state, (c) true layer-by-layer; multi-level: (d) transient to multilevel and (e) island growth. The RHEED intensities in (a) and (b) are reprinted with permission from Lipmaa *et al.* (2000).

the surface. No nucleation of 2D islands therefore occurs. As the average step density remains relatively small, the introduced step density during each laser pulse increases substantially, which is observed as a relatively large intensity modulation of the RHEED signal.

Secondly, a similar mode (see Fig. 1.6(b)), is found; however the existing steps are now formed by (pyramidal) shaped islands, which are constantly growing without any change of morphology. Therefore, nucleation has to take place, which was by definition forbidden in true step flow growth. This is confirmed by AFM in that many small islands and meandering steps are observed (Rijnders *et al.*, 2004) and is often observed for layered oxide materials (Dam and Stauble-Pumpin, 1998).

In Fig. 1.6(c) true 2D RHEED intensity oscillations are observed when $SrTiO_3$ is deposited with PLD, at a temperature of 850°C and an oxygen pressure of 3 Pa (Koster *et al.*, 1998b). Also at lower temperatures, see Fig. 1.5, this behaviour is still visible. In this figure, another feature of PLD is exemplified: the relaxation phenomenon. In the case of PLD, a typical value for the deposition rate within one pulse is of the order of $10\,\mu m/s$ (Koster *et al.*, 1998b; Geohegan and Purektzky, 1995). Therefore, high supersaturation is expected (Rijnders and Blank, 2007a) when the plume is on and thus the number of 2D nuclei can be very high. Subsequently, when the plume is off, larger islands are formed through re-crystallization, manifested as the typical relaxation of the RHEED intensity of the specular spot (Karl and Stritzker, 1992). The insets in Fig. 1.5(a) are enlargements of the relaxation after a laser pulse. From this it is clear that the characteristic relaxation time depends, among other things, on the coverage during deposition.

The relatively high temperature, in combination with a low oxygen pressure, enhances the mobility of the adatoms on the surface and, therefore, the probability of nucleation on top of a 2D island is minimized. The as-deposited adatoms can migrate to the step edges of 2D islands and nucleation takes place only on fully completed layers.

Figure 1.6(d) summarizes the response when depositing $SrTiO_3$ at the temperature of 800°C and the oxygen pressure of 10 Pa, with a continuous pulse frequency of 1 Hz. Apparently, the surface is evolving from a single-level system to a multilevel system, as indicated by the strong damping of RHEED intensity oscillations in Fig. 1.6(d). A higher pressure causes the mobility of the particles on the surface to be lower. The probability of nucleation on top of a two-dimensional island has increased, resulting in the observed roughening. In Fig. 1.6(d), AFM shows the surface after deposition of a 30 nm thick $SrTiO_3$ film. Also from this image it is clear that the surface consists of multiple levels: at least four unit-cell levels can be seen.

In Fig. 1.6(e) the observed variations of the specular intensity during deposition of $CaCuO_2$ are shown. Initially, there are two oscillation periods visible after which the intensity exponentially drops to below the detector limit. This is an example of a layer-by-layer growth mode, followed by 3D island growth, also schematically drawn in Fig. 1.6(e). This growth mode is often accompanied by a transmission diffraction pattern.

1.6.1 Pulsed laser interval deposition

To overcome the roughening at lower temperatures and higher pressures, several groups have suggested the use of periodically interrupted growth (Gupta *et al.*, 1993; Rijnders *et al.*, 1997), leading to smoother surfaces. Annealing the surface will level off any roughness that has developed during deposition. This option is especially useful in the case of co-evaporated thin films. However, considerable mobility is still needed on the surface, despite the longer waiting times. Sometimes increasing the temperature to increase the activity is an option. In the next section, a much more effective method is presented, without changing the temperature and pressure that are determined by the phase stability of the film material.

From the point of view of phase stability for many oxide materials, temperature and oxygen pressure have to be fixed. The only way to obtain true layer-by-layer growth is to apply manipulated growth, reducing the island size during growth. Here, a method will be briefly discussed, based on the high supersaturation attained during PLD; pulsed laser interval deposition. Although there is the general impression that, owing to the pulsed deposition, the growth mechanism differs partially from continuous physical and chemical deposition techniques, this has rarely been made use of. Here, we introduce a growth method based on a periodic sequence: fast deposition of the amount of material needed to complete one monolayer followed by an interval in which no deposition takes place and the film can reorganize. This makes it possible to grow in a layer-by-layer fashion in a growth regime (temperature, pressure) where otherwise island formation would dominate the growth.

Since small islands promote interlayer mass transport, one can utilize the high supersaturation achieved by PLD (Koster *et al.*, 1999; Rijnders and Blank, 2007a) by maintaining it for a longer time interval and suppressing subsequent coarsening. Accordingly, to circumvent premature nucleation due to the limited mobility of the adatoms at a given pressure and temperature, causing a multilevel 2D growth mode, we have utilized interval deposition. Exactly one unit-cell layer is deposited in a very short time interval, of the order of the characteristic relaxation times, followed by a much longer interval during which the deposited material can rearrange. During the short deposition intervals, only small islands will be formed due to the high supersaturation typical for PLD. The probability of nucleation on the islands increases with their average radius (Rijnders and Blank, 2007a; Christen and Eres, 2008) and is, therefore, small in the case of fast deposition. The total amount of pulses needed to complete one unit-cell layer has to be as high as possible, to minimize the error introduced by the fact that only an integer number of pulses can be given. Both a high deposition rate and sufficiently accurate deposition of one unit-cell layer can be obtained by PLD using a high laser pulse frequency.

To prove the validity of this growth method, SrTiO$_3$ was deposited using the KrF excimer laser with maximum repetition rate of 10 Hz as well as the XeCl excimer laser with a maximum repetition rate of 100 Hz. The same oxygen pressure and substrate temperature was used as in the case of continuous deposition where strong damping of the specular intensity is observed (Fig. 1.6(a)).

Figure 1.7 shows the RHEED intensity during 10 cycles of deposition (at 10 Hz) followed by a period of no deposition. In this case the number of pulses needed per unit-cell layer was estimated to be about 27. The decay of the intensity after each unit-cell layer is significantly lower compared with the situation in Fig. 1.6(d).

From Fig. 1.7, it can be seen that the recovery of the intensity after each deposition interval is fast at the beginning of deposition. The decrease in intensity can be ascribed to the fact that only an integer number of pulses can be given to complete a unit-cell layer, besides nucleation on the next level. A slightly lower or higher coverage causes a change in recovery time. This situation deteriorates with every subsequent unit-cell layer, as follows from increasing relaxation times.

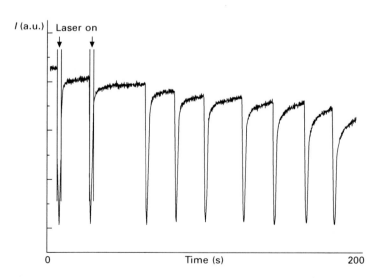

1.7 Intensity variations of the specular reflection during interval deposition of SrTiO$_3$ using a laser repetition rate of 10 Hz. Reprinted with permission from Koster *et al.* (1999).

1.7 RHEED intensity variations and Monte Carlo simulations

In order to describe the crystallization processes on the surface during thin film growth, an often-used model is the solid-on-solid (SOS) model introduced by Weeks and Gilmer (2007) and (Vvedensky et al., 1990). Diffusion of deposited material on a simple cubic crystal surface is described in terms of single-particle lattice hopping. If we assume that the particles on the oxide surface are cubes with the size of one unit-cell, the SOS model can be applied. In reality, the diffusion process is expected to be much more complicated, consisting of (de)composition steps, and the actually moving particles should be much smaller. However, effectively the process can be described by the simple cubic model. Note that the oxygen pressure will probably affect the rate of each reaction step. In our case, the oxygen pressure is assumed to be high enough and not to be rate limiting.

The diffusion kinetics is described by an Arrhenius hopping process on a $l \times l$ matrix, determined by the surface diffusion barrier E_S and the nearest-neighbour coordination n ($n = 0, 1, ..., 4$) of each particle with a lateral bond strength E_N. The hopping probability k is then given by:

$$k = k_0 \exp\left[\frac{-(E_S + nE_N)}{k_B T}\right] \quad [1.16]$$

where k_0 represents an attempt frequency (~10^{13} Hz) for atomic processes. To simulate an infinitely large crystal surface, periodic boundary conditions are applied. The Monte Carlo simulation method applied here has been extensively described by others (Vvedensky and Clarke, 1990; Maksym, 1988) and details can be found elsewhere (Van Setten, 1999).

The RHEED intensity, in view of the step density model, can be calculated as follows for the $l \times l$ matrix (Achutharaman et al., 1994):

$$L = \frac{1}{l^2} \sum_{i,j} |h_{i,j} - h_{i,j+1}|\cos\phi + |h_{i,j} - h_{i+1,j}|\sin\phi \quad [1.17]$$

and

$$I_{RHEED} \propto 1 - \sigma \quad [1.18]$$

where σ is the step density, $h_{i,j}$ are step heights on the surface and ϕ the azimuthal angle of the electron beam with respect to the principal directions of the matrix.

The pulsed way of deposition is simulated here by depositing a number of particles on the surface randomly and instantaneously, after which the particles can diffuse, when the next pulse arrives is determined by the repetition rate of the laser.

Figure 1.8 shows the simulation of continuous deposition using $E_S = E_A$

24   In situ characterization of thin film growth

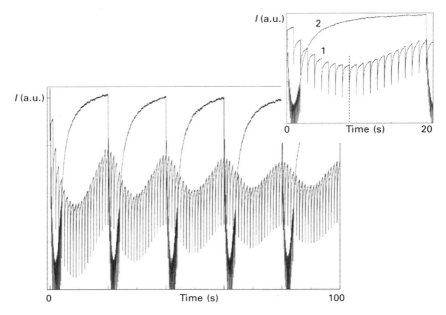

1.8 Simulated intensity variations during continuous deposition at 815 °C using a laser repetition rate of 1 Hz (curve 1) and interval deposition at the same temperature using a laser repetition rate of 10 Hz (curve 2). The deposition rate was set at 20 pulses per unit-cell layer. The inset shows the first unit-cell layer.

$= 2.2\,\text{eV}$, $E_N = 0.27\,\text{eV}$, $k_0 = 10^{13}\,\text{Hz}$, $f = 0°$ and $T = 815\,°\text{C}$ (curve 1). For these settings we observe damped intensity oscillations with a period of 20 seconds, indicating roughening of the surface. The relaxation behaviour of the intensity after each laser pulse, as seen in the inset of Fig. 1.8 (curve 1), depends on the coverage. These observations are in agreement with the experiments. Also the characteristic times obtained from the simulations agree with the measured values. Because the sampling rate (number of data points) is much higher than in the experiments, the relaxation behaviour is very pronounced even at the minimum, and the characteristic times can still be determined. Note that the amplitude decreases over time due to surface roughening. Finally, for the first period the minimum intensity seems to be shifted toward $\theta < 0.5$, as indicated by the dashed line in the inset of Fig. 1.8, which is also observed in experiments.

When interval deposition is simulated, applying a laser repetition rate of 10 Hz, the typical shape of interval deposition emerges, as seen in Fig. 1.8 (curve 2). From these simulations it is clear that after each unit-cell layer, the intensity remains much higher in the case of interval deposition, which was also observed in the experiments.

1.8 Conclusions

In situ RHEED intensity monitoring of the specular reflection gives information about the nucleation of the deposited material. Moreover, by analysing the RHEED pattern, one can judge the surface quality and detect for example the origination of precipitates.

The combination of the pulsed method of deposition with intensity monitoring results in the possibility of time-resolved RHEED. Besides the observed intensity oscillations in the case of layer-by-layer growth, which enables accurate growth rate control, it is clear that intensity relaxation observed due to the typical pulsed method of deposition leads to a wealth of information about growth parameters.

To apply RHEED for growth rate monitoring at an atomic level, it is necessary to have a 2D or layer-by-layer growth mode. If possible, the deposition parameters should be adjusted to obtain a layer-by-layer growth mode. Yet, in view of phase stability, many oxide materials are not stable over large enough temperature and pressure ranges. The growth modes of layered oxide materials and the use of growth manipulation to impose a layer-by-layer growth mode have been discussed.

For successive deposition of unit-cell layers of different targets to create artificially layered structures, however, a true layer-by-layer mode is essential throughout the deposition of the entire film, since for every layer the starting surface has to be atomically flat again. A transition to multilevel has to be avoided as much as possible because the growth front may end up being thicker than each constituting layer.

For many oxide materials one cannot freely choose the deposition temperature and pressure, as in the case of homoepitaxy, to obtain layer-by-layer growth, in view of the phase stability. Often, the desired phase is stable only in a limited regime of the deposition conditions, depending on the substrate material.

The mechanisms resulting in the typical relaxation behaviour of the specular intensity after each deposition interval are expected to be more complex than in relaxation during continuous layer-by-layer growth, i.e. they do not consist of diffusion of non-interacting particles only. Therefore, we used the solid-on-solid model and simulated interval deposition using the Monte Carlo method. Experimental parameters determined earlier were used to characterize homoepitaxy of $SrTiO_3$.

As an example, a clear difference in the calculated RHEED intensity variations is observed during continuous growth and interval deposition, which resemble the experimental results, both qualitatively as well as quantitatively. By comparing the characteristic relaxation times found for simulations using different laser repetition rates, a critical value of this repetition rate could be determined for the interval deposition method to be most effective; the deposition interval has to be shorter than this characteristic time.

1.9 Acknowledgements

The author would like to acknowledge Guus Rijnders, Dave H. A. Blank and Rik Groenen for discussions and help with graphics.

1.10 References

Achutharaman, V.S., N. Chandrasekhar, O.T. Valls and A.M. Goldman, Origin of RHEED intensity oscillations during the growth of $(Y,Dy)Ba_2Cu_3O_{7-x}$ thin-films, *Phys. Rev. B* **50**, 8122 (1994)

Arrott, A.S., Introduction to reflection high energy electron diffraction, in *Ultrathin Magnetic Structures I*, J.A.C. Bland and B. Heinrich, Eds, Springer-Verlag, Berlin, Heidelberg (1994), pp. 177–219

Barabasi, A.L., *Fractal Concepts in Surface Growth*, Cambridge University Press, Cambridge, UK (1995)

Beeby, J.L., in *Reflection High-energy Electron Diffraction and Reflection Electron Imaging of Surfaces*, P.K. Larsen and P.J. Dobson, Eds, Plenum Press, London (1989), pp. 29–42

Blank, D.H.A., A.J.H.M. Rijnders, G. Koster and H. Rogalla, *In-situ* monitoring during pulsed laser deposition using RHEED at high pressure, *Appl. Surf. Sci.* **127**, 633 (1998)

Blank, D.H.A., G.J.H.M. Rijnders, G. Koster and H. Rogalla, *In-situ* monitoring by reflective high energy electron diffraction during pulsed laser deposition, *Appl. Surf. Sci.* **138**, 17 (1999)

Christen, D. and Eres, G., Recent advances in pulsed-laser deposition of complex oxides, *J. Phys. – Condensed Matter* **20**, 264005 (2008)

Dam, B. and B. Stauble-Pumpin, Growth mode issues in epitaxy of complex oxide thin films, *J. Mater. Sci.-Mater. El.* **9**, 217 (1998)

Dudarev, S.L., D.D. Vvedensky and M.J. Whelan, Dynamical electron-scattering from growing surfaces, *Surf. Sci.* **324**, L355 (1995)

Geohegan, D.B. and Purektzky, A.A., Dynamics of laser-ablation plume penetration through low-pressure background gases, *Appl. Phys. Lett.* **67**, 197 (1995)

Gupta, A., M.Y. Chern and B.W. Hussey, Layer-by-layer growth of cuprate thin-films by pulsed laser deposition, *Physica C* **209**, 175 (1993)

Heyn, Ch., T. Franke, R. Anton and M. Harsdorff, Correlation between island-formation kinetics, surface roughening, and RHEED oscillation damping during GaAs homoepitaxy, *Phys. Rev. B* **56**, 13483 (1997)

Hirata, K., F. Baudenbacher, H.P. Lang, H.-J. Güntherodt and H. Kinder, Scanning-tunneling-microscopy study on $YBa_2Cu_3O_{7-x}$ films with growth stop by observing reflection high-energy electron-diffraction oscillations, *J. Alloy Compd.* **195**, 105 (1993)

Jorritsma, L., Growth anisotropy in Cu(001) homoepitaxy, PhD thesis, University of Twente (1997)

Karl, H., *In-situ* RHEED studies of YBCO-film growth during pulsed laser deposition, *IEEE Trans. Appl. Supercond.* **3**, 1594 (1993)

Karl, H. and B. Stritzker, Reflection high-energy electron-diffraction oscillations modulated by laser-pulse deposited $YBa_2Cu_3O_{7-x}$, *Phys. Rev. Lett.* **69**, 2939 (1992)

Kawamura, T., in *Reflection High Energy Electron Diffraction and Reflection Electron Imaging of Surfaces*, P.K. Larsen and P.J. Dobson, Eds, Plenum Press, London (1989), pp. 501–522

Korte, U. and P.A. Maksym, Role of the step density in reflection high-energy electron diffraction: questioning the step density model, *Phys. Rev. Lett.* **78**, 2381 (1997)

Koster, G., A.J.H.M. Rijnders, D.H.A. Blank and H. Rogalla, *Mater. Res. Soc. Symp. Proceedings* **526**, 33 (1998a)

Koster, G., B.L. Kropman, A.J.H.M. Rijnders, D.H.A. Blank and H. Rogalla, Quasi-ideal strontium titanate crystal surfaces through formation of strontium hydroxide, *Appl. Phys. Lett.* **73**, 2920 (1998b)

Koster G., G. Rijnders, D.H.A. Blank and H. Rogalla, Imposed layer-by-layer growth by pulsed laser interval deposition, *Appl. Phys. Lett.* **74**, 3729(1999)

Lagally, M.G., D.E. Savage and M.C. Tringides, in *Reflection High Energy Electron Diffraction and Reflection Electron Imaging of Surfaces*, P.K. Larsen and P.J. Dobson, Eds, Plenum Press, London (1989), pp. 139–174

Lehmpfuhl, G., A. Ichimiya and H. Nakahara, Interpretation of RHEED oscillations during MBE growth, *Surf. Sci. Lett.* **245**, L159 (1991)

Lent, C.S. and P.I. Cohen, Diffraction from stepped surfaces. 1. Reversible surfaces, *Surf. Sci.* **139**, 121 (1984)

Lipmaa, M., N. Nakagawa, M. Kawasaki, S. Ohashi and H. Koinuma, Growth mode mapping of $SrTiO_3$ epitaxy, *Appl. Phys. Lett.* **76**, 2439 (2000)

Maksym, P.A., Fast monte-carlo simulation of MBE growth, *Semicond. Sci. Technol.* **3**, 594 (1988)

Markov, I.V., *Crystal Growth for Beginners*, World Scientific, London (1995), pp. 81–86

Mitura, Z., S.L. Dudarev and M.J. Whelan, Phase of RHEED oscillations, *Phys. Rev. B* **57**, 6309 (1998)

Mutaftschiev, B., in *Kinetics of Ordering and Growth at Surface*, M.G. Lagally, ed., Plenum Press, London (1990), pp. 169–188

Myers-Beahton, A.K. and D.D. Vvedensky, Nonlinear equation for diffusion and adatom interactions during epitaxial-growth on vicinal surfaces, *Phys. Rev. B* **42**, 5544 (1990)

Myers-Beahton, A.K. and D.D. Vvedensky, Generalized Burton–Cabrera–Frank theory for growth and equilibration on stepped surfaces, *Phys. Rev. A* **44**, 2457 (1991)

Neave, J.H., B.A. Joyce, P.J. Dobson and N. Norton, Dynamics of film growth of GaAs by MBE from RHEED observations, *Appl. Phys. A* **31**, 1 (1983)

Neave, J.H., P.J. Dobson, B.A. Joyce and J. Zhang, Reflection high-energy electron-diffraction oscillations from vicinal surfaces – a new approach to surface-diffusion measurements, *Appl. Phys. Lett.* **47**, 100 (1985)

Nishikawa, H., M. Kanai and T. Kawai, Heteroepitaxy of perovskite-type oxides on oxygen-annealed $SrTiO_3$(1 0 0) – important factors for preparation of atomically flat oxide thin films, *J. Crystal Growth* **179**, 467 (1997)

Papajova, D. and M. Vesely, Model of the RHEED intensity oscillations based on reflectivity of the MBE grown surface, *Vacuum* **49**, 297 (1998)

Pukite, P.R., P.I. Cohen and S. Batra, in *Reflection High Energy Electron Diffraction and Reflection Electron Imaging of Surfaces*, P.K. Larsen and P.J. Dobson, Eds, Plenum Press, London (1989), pp. 427–447

Ready, J.F., Development of plume of material vaporized by giant-pulse laser, *Appl. Phys. Lett.* **3**, 11 (1963)

Rijnders, A.J.H.M. and D.H.A. Blank, Growth kinetics during PLD, in *Pulsed Laser Deposition of Thin Films: Applications-led Growth of Functional Materials*, R. Eason, ed., Wiley (2007a), pp. 177–190.

Rijnders, G. and D.H.A. Blank, Materials science – build your own superlattice, *Nature* **433**, 369 (2007b)

Rijnders, A.J.H.M., G. Koster, D.H.A. Blank and H. Rogalla, *In situ* monitoring during pulsed laser deposition of complex oxides using reflection high energy electron diffraction under high oxygen pressure, *Appl. Phys. Lett.* **70**, 1888 (1997)

Rijnders, G., D. Blank, J. Choi and C. Eom. Enhanced surface diffusion through termination conversion during epitaxial $SrRuO_3$ growth, *Appl. Phys. Lett.* **84**, 505 (2004)

Rosenfeld, G., B. Poelsema and G. Comsa, Epitaxial growth modes far from equilibrium, in *Growth and Properties of Ultrathin Epitaxial Layers*, D.A. King and D.P. Woodruff, Eds, Elsevier Science BV (1997), pp. 66–101

Sakamoto, T., N.J. Kawai, T. Nakagawa, K.Ohta, T. Kojima and G. Hashiguchi, RHEED intensity oscillations during silicon MBE growth, *Surf. Sci.* **174**, 651 (1986)

Shitara, T., D.D. Vvedensky, M.R. Wilby, J. Zhang, J.H. Neave and B.A. Joyce, Misorientation dependence of epitaxial-growth on vicinal GaAs(001), *Phys. Rev. B* **46**, 6825 (1992a)

Shitara, T., J. Zhang, J.H. Neave and B.A. Joyce, Ga adatom incorporation kinetics at steps on vicinal GaAs(001) surfaces during growth of GaAs by molecular-beam epitaxy, *J. Appl. Phys.* **71**, 4299 (1992b)

Shitara, T., D.D. Vvedensky, M.R. Wilby, J. Zhang, J.H. Neave and B.A. Joyce, Step-density variations and reflection high-energy electron-diffraction intensity oscillations during epitaxial-growth on vicinal GaAs(001), *Phys. Rev. B* **46**, 6815 (1992c)

Stoyanov, S. and M. Michailov, Non-steady state effects in MBE – oscillations of the step density at the crystal-surface, *Surf. Sci.* **202**, 109 (1988)

Terashima, T., Y. Bando, K.Iijima, K. Yamamoto, K. Hirata, K. Hayashi, K. Kamigaki and H. Terauchi, Reflection high-energy electron-diffraction oscillations during epitaxial-growth of high-temperature superconducting oxides, *Phys. Rev. Lett.* **65**, 2684 (1990)

Van der Wagt, J.P.A., Reflection high-energy electron diffraction during molecular-beam epitaxy, PhD thesis, Stanford University, USA (1994)

Van Hove, J.M., C.S. Lent, P.R. Pukite and P.I. Cohen, Damped oscillations in reflection high-energy electron-diffraction during GaAs MBE, *J. Vac. Sci. Technol. B* **1**, 741(1983)

Van Setten, E., M.Sc. thesis, University of Twente (1999)

Vvedensky, D.D. and Clarke, S., Recovery kinetics during interrupted epitaxial-growth, *Surf. Sci.* **225**, 373 (1990)

Vvedensky, D.D., S. Clarke, K.J. Hugill, A.K. Myers-Beaghton and M.R. Wilby, in *Kinetics of Ordering and Growth at Surface*, M.G. Lagally, ed., Plenum Press, London (1990), pp. 297–311

Weeks, J.D. and J.H. Gilmer, Dynamics of crystal growth, *Adv. Chem. Phys.* **40**, 157 (2007)

Yang, H.-N., *Diffraction from Rough Surfaces and Dynamic Growth Fronts*, World Scientific (1993), chapter 2

Yang, H.-N, G.-C. Wang and T.-M. Lu, Quantitative study of the decay of intensity oscillations in transient layer-by-layer growth, *Phys. Rev. B* **51**, 17932 (1995)

Zandvliet, H.J.W., H.B. Elswijk, D. Dijkkamp, E.J. van Loenen and J. Dieleman, On the period of reflection high-energy electron diffraction intensity oscillations during Si molecular-beam epitaxy on vicinal Si(001), *J. Appl. Phys.* **70**, 2614 (1991)

Zhang, J., J.H. Neave, P.J. Dobson and B.A. Joyce, Effects of diffraction conditions and processes on RHEED intensity oscillations during the MBE growth of GaAs, *Appl. Phys. A* **42**, 317 (1987)

2
Inelastic scattering techniques for *in situ* characterization of thin film growth: backscatter Kikuchi diffraction

N. J. C. INGLE, University of British Columbia, Canada

Abstract: The inherent characteristics of electron diffraction are particularly useful for determining the full structure (lattice parameters, space group, atomic positions) of thin films *in situ* and in real-time. Electrons are easy to generate, manipulate, and detect, and they have a strong interaction with matter. However, the dynamic nature of electron diffraction generally makes detailed analysis complicated. Along with the multiple scattering complications comes one significant benefit: the formation of Kikuchi diffraction patterns. These contain within them a representation of the full three-dimensional structure of the probed material.

Key words: inelastic scattering, backscatter Kikuchi diffraction, thin film structural determination, dynamic electron scattering, electron backscatter diffraction.

2.1 Introduction

The geometrical constraints within a molecular beam epitaxy (MBE)-like deposition chamber, and the small quantity of material available for study, make determining the full structure (lattice parameters, space group, atomic positions) of a thin film *in situ* and in real-time a serious challenge. The inherent characteristics of electron diffraction are particularly useful for approaching this problem. Electrons are easy to generate, manipulate, and detect, and they have a strong interaction with matter. However, the dynamic nature of electron diffraction generally make its detailed analysis complicated. There is a significant benefit to the multiple scattering complications: the formation of Kikuchi diffraction patterns, which contain a representation of the full three-dimensional structure of the probed material.

Kikuchi diffraction patterns are the result of diffraction from inelastically scattered electrons, and they can be found when using all electron diffraction techniques. They are used most widely in transmission electron microscopy (TEM) and scanning electron microscopy (SEM) as techniques for tracking the orientation of a crystal structure in space. It is convenient to think of these Kikuchi patterns as being generated from a point source of electrons located within the diffracting medium; and doing so provides a framework for understanding their simultaneous access to all of the diffraction planes in a sample. This leads to data-intensive, complex, and rich diffraction patterns. The access to all of a sample's diffraction planes, and the ease of collecting this type of diffraction pattern, makes it a potentially invaluable tool for the *in situ* and real-time characterization of thin films.

2.2 Kikuchi patterns

In the 1920s, while studying thin mica films using an electron beam, S. Kikuchi (1928) observed a background structure, as well as the expected diffraction peaks, up to 20° away from the direction of the beam. This background structure consisted of a series of parallel line pairs, later called Kikuchi lines. Kikuchi developed a fairly straightforward interpretation of these lines, often referred to as the 'two-event model'. In this model the incoming collimated and mono-energetic beam of electrons are diffused in the crystal by an unspecified scattering process. In essence, this first step generates a source of electrons traveling in multiple directions, inside the material. The second step is then standard Bragg diffraction of the diffused electrons from the planes in the material.

From any point inside the crystal there are two possible angles for which Bragg diffraction from a single set of lattice planes can occur. Therefore, each set of crystallographic planes generates two diffraction cones with a large opening angle of $180° - 2\theta_{hkl}$ and an axis perpendicular to the diffracting lattice planes. These cones will be seen on a collecting screen as a pair of lines, i.e. the Kikuchi lines (see Fig. 2.1). These lines will be parallel to,

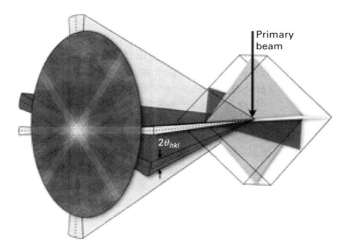

2.1 A schematic of the Kikuchi band formation highlighting the Kikuchi band's connection to the lattice planes.

and on either side of, the intersection of the diffracting lattice plane with the screen. In reality, the multiple scattering process required to generate Kikuchi patterns leads to more complex contrasts in the image than simple pairs of lines (Reimer, 1985). The diffraction pattern is better described as bands with excess or deficient Kikuchi lines that define the edge of the band.

Several years after Kikuchi had observed Kikuchi patterns in transmission, transmission Kikuchi patterns (TKPs), the same types of patterns – consisting of parallel line pairs – were observed up to 160° from the incoming beam of electrons (von Meibom and Rupp, 1933; Boersch, 1937). These patterns are now described as backscatter Kikuchi patterns (BKP). Although the phenomenological 'two-event model' description of the formation of the TKP and BKP are similar and the geometrical interpretation of the patterns are identical, there has been debate as to whether the same mechanisms are involved in TKP and BKP formation due to the large difference in scattering angles (Alam *et al.*, 1954; Reimer, 1985; Wells, 1999; Zaefferer, 2007).

Simulations that use thermal diffuse scattering as the primary process of incoherent and inelastic scattering agree quite well with experimental TKPs (Omoto *et al.*, 2002). Recently it has also been argued that because of the lower primary beam energies used to generate BKPs, the same thermal diffuse scattering can also account for the large scattering angles in BKP formation (Zaefferer, 2007).

2.2.1 TEM

TEM relies heavily on TKP patterns to help determine the orientation and tilt axis of a sample in the microscope. This is primarily due to the fact that the Kikuchi patterns are an image of the real crystal symmetry and therefore behave as if attached to the crystal, i.e. they provide a continuous pattern that moves in the same sense as the crystal. This allows a Kikuchi map to be used as a roadmap for moving around reciprocal space. This can be advantageous for stereomicroscopy, contrast work, and selected area diffraction alignment (Gareth, 1979).

TKPs also provide a means for general structure analysis in TEM, including structure and phase identification, grain boundary mapping, and texture mapping. Convergent beam electron diffraction, generally referred to as CBED, provides access to the symmetry of the crystal and to the position of atoms within the unit cell, on crystals down to about 2 nm in size. CBED is inherently a dynamic diffraction problem, and as such the presence of Kikuchi bands within the diffraction pattern are a significant part of the CBED pattern analysis (Omoto *et al.*, 2002).

2.2.2 SEM

There has been a great deal of interest in BKP within the materials science community because of its usefulness in studying texture and orientation relationships on a micron length scale in SEM (see, for example, Venables *et al.*, 1976; Dingley *et al.*, 1989; Dingley and Randle, 1992; Randle, 1992; Field, 1997; Dingley, 2004). In this style of work, the electron beam is scanned over a sample and BKPs are collected at each point in the scan. Crystallographic orientational changes from one area in the sample to the next are then easily identified as a change in the BKP, thereby generating a map of the grain structure. This technique is normally called electron backscatter diffraction (EBSD). In addition to this work on orientational mapping, EBSD patterns collected from an SEM have also been used for phase identification by crystallographic point group or space group classification, even on nanoparticles (Baba-Kishi and Dingley, 1989a, 1989b; Dingley *et al.*, 1995; Small *et al.*, 2002), and the evaluation of both plastic and elastic strains (Troost *et al.*, 1993; Wilkinson, 1996).

The angular dependent variations seen in the backscatter electron yield in SEM, called electron channeling patterns (ECPs), are very similar to the EBSD patterns, and it has been suggested that the ECPs and EBSD patterns are theoretically related by the reciprocity theorem (Venables and Harland, 1973; Reimer, 1985).

2.2.3 Kikuchi pattern analysis

A Kikuchi pattern is, in essence, a gnomonic projection of the intersection of all the sample's lattice planes with a sphere of reflection centered on the electron source point (see Fig. 2.1). As such, the intersection of different Kikuchi lines indicates the location of a zone axis, and displays the symmetry of the zone axis. Based on these ideas, general procedures have been developed for using a BKP (or TKP) to determine the unit cell and space group of a sample (Baba-Kishi and Dingley, 1989a, 1989b; Dingley *et al.*, 1995; Dingley and Wright, 2009).

The analysis of the BKP to obtain space group information starts with the analysis of individual zone axes (i.e. points where at least two sets of Kikuchi lines cross), in order to define their respective point group symmetries (see Fig. 2.2). Since the BKP does not distinguish between point groups with and without the inversion symmetry, there are only 11 possible symmetries to be considered. These symmetries are the so-called Laue groups. The combination of the Laue groups from several zone axes will then allow the determination of the crystallographic point group (see Table I from Baba-Kishi and Dingley, 1989a, or Dingley *et al.*, 1995). From this point on, determining the space group requires the ability to calculate some rough lattice parameters from the

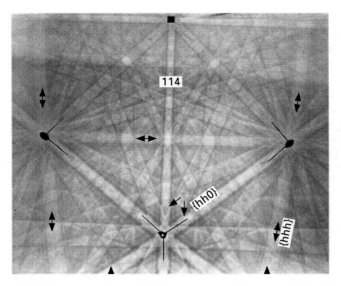

2.2 BKP from a cleaved (001) crystal of silicon at 40 kV with crystallographic symmetry marked on several zone axes. Reprinted with permission (Baba-Kishi, 2002).

ratios of the d-spacings and interzonal angles of the principal reflections, then simulate the BKP, and finally test the resulting match to the experimental pattern. With some *a priori* knowledge of the Bravais lattice, this procedure can be fully automated with band detection algorithms based on the Hough transformation (Lassen *et al.*, 1992; Schwarzer, 1997). This automation is possible using most commercial EBSD software packages.

It is worth highlighting that Kikuchi pattern formation is inherently a multiple scattering phenomenon, which can influence the intensity comparison between the experimental patterns and simulated patterns based on kinematic theory.

2.3 Kikuchi lines in reflection high-energy electron diffraction (RHEED) images

In addition to coherent Bragg diffraction, reflection high-energy electron diffraction (RHEED) images generally also include a diffuse background, sharp Kikuchi lines, and surface resonance features (see Fig. 2.3a). These additional features are generally spectroscopically separated from the coherent Bragg diffraction peaks (Horio *et al.*, 1996; Staib *et al.*, 1999; Nakahara *et al.*, 2003), and energy filtering can remove them.

The Kikuchi patterns found in RHEED images can be quite sharp and contain significant intensity in the case of crystals with very high bulk and surface order. As with TKP in the TEM, the Kikuchi patterns seen on the RHEED screen are easily distinguishable from the coherent Bragg diffraction intensity because they move in a continuous and coincident manner when the crystal is rotated. Because of this, they can be used to align the azimuthal

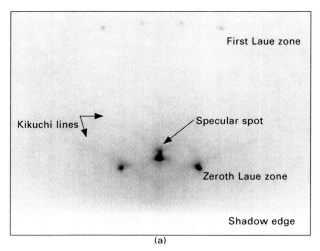

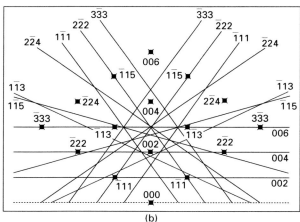

2.3 (a) An exemplary RHEED image with the shadow edge, specular spot, Laue zones and Kikuchi lines identified. Reprinted with permission (Ingle et al., 2010). (b) Theoretical Kikuchi line pattern for the crystal surface geometry of a face centred cubic (fcc) (001) surface along the [110] direction. No surface potential corrections are included. Reprinted with permission (Braun et al., 1998b). (c) Kikuchi line fits to the diffraction patterns of GaAs (001) β(2 × 4) reconstructed surface, which include the surface potential. The arrows indicate the most reliable points for fitting. Reprinted with permission (Braun et al., 1998a).

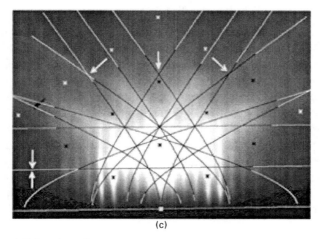

2.3 Continued.

direction of the sample with respect to the incoming electron beam with high accuracy.

A simple reciprocal lattice-based method can be used to describe the expected locations of Kikuchi lines on the RHEED screen (Dobson, 1988; Gajdardziska-Josifovska and Cowley, 1991; Braun, 1999; Ichimiya and Cohen, 2004). After the first step of the two-event model has occurred – the inelastic scattering step – electrons will be traveling in all directions within the crystal. These electrons then scatter, with conservation of energy and momentum, in the second step. This requires that both the incident (k_0) and the scattered wavevector (k) for the second step of the two-event model have equal magnitudes and that the scattering vector ($k - k_0$) be a vector in the reciprocal lattice (g). Mathematically this can be written as $2k_0 \cdot g + g^2 = 0$, which when drawn for all possible k_0 vectors produces the Brillouin zone boundaries for the given reciprocal lattice. Therefore, to identify the Kikuchi patterns seen in the RHEED image for any particular azimuthal direction, we must first plot the reciprocal lattice points perpendicular to that direction, and then the perpendicular bisectors of those reciprocal lattice points, i.e. the Brillouin zone boundaries, can be drawn (see Fig. 2.3(b)). This will produce all the possible geometries of the Kikuchi lines, although far fewer lines will be seen in practice.

The grazing angles of the scattered electrons that form the Kikuchi patterns in the RHEED image lead to several significant effects. The first is that the sample's surface potential strongly alters the perpendicular momentum of the scattered electrons in this geometry, causing the Kikuchi lines to become curved (see Fig. 2.3(c)). This refraction effect has been used by Braun (Braun *et al.*, 1998a; Braun, 1999) to quantitatively determine the surface potential needed for further dynamic scattering calculations. Braun was also able to extract the mis-orientation of a sample's surface with respect to the underlying crystallographic orientation (i.e. the sample miscut) from fitting the RHEED-based Kikuchi patterns (Braun, 1999). In addition, the presence of terraces on a surface can change the behavior of the refraction depending on whether an electron leaves the surface through the edge of a terrace, or through the terrace surface (Dobson, 1988). The final effect is that the electrons that generate these grazing angle Kikuchi patterns are expected to originate primarily from the near surface region, and therefore may not generate Kikuchi patterns that are representative of the bulk. Kikuchi patterns for two-dimensional structures have been determined and evidence found in the annealed and reconstructed MgO (111) surface (Gajdardziska-Josifovska and Cowley, 1991) and a vacuum cleaved GaAs (110) surface (Braun, 1999).

One additional complication associated with the presence of Kikuchi patterns in the RHEED image is that there can be significantly enhanced intensity where the Kikuchi patterns and the coherent Bragg diffraction patterns overlap. This effect is called a surface-wave resonance (Wang, 1996), and can lead to an order of magnitude increase in the intensity at these locations.

Baba-Kishi (1990) collected RHEED images within an SEM, allowing a geometry and beam energy that could collect electrons over a much wider angular range than is standard for RHEED measurements. Figure 2.4 shows the RHEED pattern collected at 40 keV from a (100) surface of GaAs. Close to the pattern center, where the contrast has been adjusted to allow features to be seen, the white arrows indicate enhanced intensity from the surface-wave resonance where the Kikuchi lines overlap with the coherent Bragg diffraction. Close to the white arrows, the Kikuchi patterns that are commonly found in RHEED images are clearly evident. Moving radially out from the center of the pattern, it is possible to observe that the Kikuchi lines normally seen in the RHEED images are the excess Kikuchi lines on one side of the Kikuchi bands that are present in BKP and TKPs. On stepping further out, there is a sharp continuous circle of intensity which is called a high-order Laue zone (HOLZ) ring. The excess Kikuchi lines which form the envelope of the HOLZ rings become bright when the deficient Kikuchi lines are close to the zone axis. The diameter of HOLZ rings is related to the distance between atoms in a zone-axis direction, and has been used by Michael and Eades (2000) to determine lattice parameters. Beyond the HOLZ rings, the Kikuchi bands become the more familiar BKPs.

The combination of intensity from both coherent and incoherent scattering processes in the RHEED image complicates quantitative analysis because it requires a unified treatment within the dynamical high-energy diffraction theory (Winkelmann, 2010). This problem has been studied for CBED because coherent diffraction, HOLZ rings, and Kikuchi patterns are simultaneously

Inelastic scattering techniques for *in situ* characterization

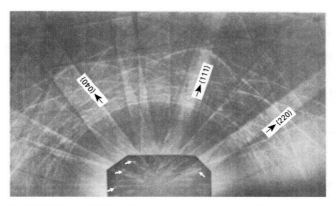

2.4 RHEED and Kikuchi pattern obtained at 40 keV from a (100) surface of single-crystal GaAs. Reprinted with permission (Baba-Kishi, 1990).

present (Cowley, 1995; Omoto *et al.*, 2002). Collecting Kikuchi patterns away from coherent scattering allows for greater ease of interpretation and modeling.

2.4 Dual-screen RHEED and Kikuchi pattern collection

Kikuchi patterns, generated by the RHEED electron gun, are present in the full hemisphere above the sample surface. Horio (2006) investigated the combination of RHEED and Kikuchi patterns on a Si(111)-7×7 surface using a conventional RHEED set-up and a second hemispherical screen in line with the sample normal. This hemispherical screen (see Fig. 2.5(a)), which led to the name astrodome RHEED, was used to enable an acceptance angle of 95°; it also minimized the distortion of Kikuchi patterns created by their standard gnomonic projection on a flat screen. Day's work (Day, 2008) provides an in-depth discussion of procedures for the collection and analysis of Kikuchi patterns from spherical surfaces.

The particular location of Horio's hemispherical screen does not allow for the use of this screen during film growth. Figure 2.5(b) shows a slightly altered geometry, with a second flat screen, that allows the Kikuchi screen to be monitored throughout the film growth processes – thereby providing *in situ* and real-time data collection. The electron beam has an angle of incidence on the sample of about 2°, as is standard for RHEED. The Kikuchi screen is 5 cm away from the sample with an angle of 45° to the sample normal. Figures 2.5(c) and (d) show simultaneous RHEED and Kikuchi images collected from a (001) $SrTiO_3$ substrate after being heated to 650 °C under atomic oxygen. With this set-up, the strongest Kikuchi band contrast occurs

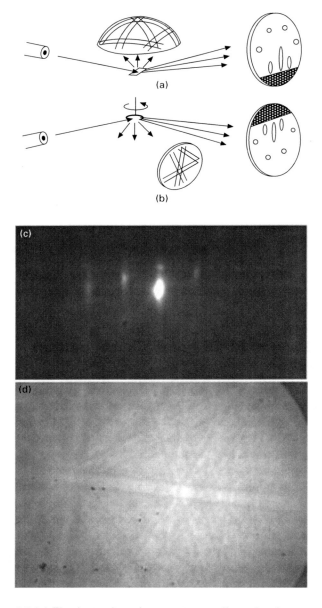

2.5 (a) The 'astrodome' geometry to allow simultaneous RHEED and Kikuchi pattern collection. Reprinted with permission (Horio, 2006). (b) A dual screen geometry to allow RHEED and Kikuchi pattern collection during film growth. The simultaneous (c) RHEED pattern and (d) Kikuchi pattern of a $SrTiO_3$ substrate at 650 °C and under a 1×10^{15} atoms/cm^2 s flux of atomic oxygen using the dual screen geometry of (b) and a 30 keV RHEED beam energy.

with an incoming electron beam incident angle of around 15°. Zaefferer (2007) noted that maximum electron intensity is obtained when the angle of incidence of the electron beam and the angle of emission of the Kikuchi patterns are similar. This suggests that it would be advantageous to place a screen to capture the Kikuchi patterns with an angle to the sample normal of as close to 90° as possible, while still avoiding the main Bragg diffraction peaks. However, as will be discussed later, the best location for the screen depends on which particular sets of Kikuchi bands provide access to the most significant structural information.

One major concern with using the Kikuchi patterns for the structural analysis of thin films is the influence of diffraction from the substrate on the observed pattern. If the film is too thin, the electron beam will interact with the substrate as well as with the film. In general, 30 keV electrons have a penetration depth of roughly 500 Å. This is then decreased to about 50 Å because of the low angle of incidence of the electron beam to the sample. In agreement with this, C.J. Harland *et al.* (1981) determined an information depth of ≤100 Å, while Kohl (1978) calculated an information depth of about 50–60 Å for the reciprocally related electron channeling patterns. This approximate probe depth of 50–100 Å is further experimentally supported by Baba-Kishi and Dingley (1989b) in work on bulk NiS_2. These results imply that a film thickness of >200 Å will be adequate to avoid the influence of the substrate on the Kikuchi pattern formation.

To gain qualitative information from the Kikuchi patterns requires two important pieces of geometrical information: the location of the pattern center – the point in the Kikuchi pattern that has no distortion from the gnomonic projection, and therefore relates back to the point source of electrons located within the sample – and the distance between the screen and the point source of electrons located within the sample. Three methods for obtaining this information are outlined in the literature: conic fitting (Biggin and Dingley, 1977; Venables and Binjaya, 1977), the circular mask technique (Venables and Binjaya, 1977), and the known orientation method (Baba-Kishi, 1998). Because epitaxial growth primarily occurs on a well-characterized single crystal substrate, 'known orientation' is the easiest and most accurate method.

In situ Kikuchi diffraction of epitaxial thin films allows several simplifications over the general procedure for analyzing the images. The first is that there is normally prior knowledge of the expected structure, so that a full *ab-initio* structure determination is not needed. The second is that the diffraction pattern of the substrate, when collected immediately prior to growth, can be directly compared with the as-grown film to determine changes in structure and to define the exact geometry. Finally, azimuthal rotation of the sample while collecting RHEED data also allows collection of a large portion of the Kikuchi pattern generated in the hemisphere above the sample, thus allowing many more zone axes to be seen. With the experimental geometry shown in Fig. 2.5(b), two-thirds of the hemisphere is visible, although owing to the flat nature of the phosphorus screen the edges of the images will be geometrically distorted.

Figure 2.6(a) shows a Kikuchi pattern, collected simultaneously with a RHEED image using the dual screen set-up of Fig. 2.5(b), of a $SrRuO_3$ film grown on $SrTiO_3$. This was collected right at the end of film growth while the substrate was still at 650 °C and exposed to an atomic oxygen flux of 1 × 10^{15} atoms/cm^2 s. Figure 2.6(b) shows a simulated pattern of tetragonal $SrRuO_3$, with various zone axes indexed, and exhibiting the same orientation as that of the experimental image.

2.5 Lattice parameter determination

Lattice parameters can be determined from Kikuchi patterns by directly measuring the width and ratios of the low-order Kikuchi lines. To calculate a d-spacing, half the angle subtended by the Kikuchi bandwidth can be plugged into the Bragg equation. However, the band edge is not very sharp: in the order of 1 mrad, as compared with the 20 mrad Kikuchi band width for 30 keV electrons. This only allows an accuracy in the 3% range (Baba-Kishi and Dingley, 1989b; Dingley et al., 1995; Wilkinson, 1996) for lattice parameters.

The HOLZ rings found in many BKPs can also be used to help determine lattice parameters. Measuring the HOLZ ring diameter allows the d-spacing of the lattice planes perpendicular to the zone axis around which the ring is

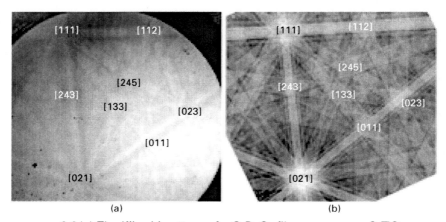

2.6 (a) The Kikuchi pattern of a $SrRuO_3$ film grown on a $SrTiO_3$ substrate at 650 °C and under a 1 × 10^{15} atoms/cm^2 s flux of atomic oxygen; zone axes are indicated. (b) The calculated Kikuchi pattern for tetragonal $SrRuO_3$ with the same orientation. Reprinted with permission (Ingle et al., 2010).

found to be calculated (Champness, 2001). This technique is more commonly used in CBED measurements. In general the HOLZ rings are sharper than the edges of Kikuchi bands; but they are not always present in BKPs, and even when they are the gnomonic projection needs to be properly accounted for. Michael and Eades (2000) used this technique and were able to obtain lattice parameter accuracies of better than 1% for many samples. However, they found that the largest error was in defining the radius of the HOLZ rings, and that error became quite large (up to 10%) for high atomic number samples.

2.6 Epitaxial film strain determination

Although determining lattice parameters from Kikuchi patterns is not very accurate, determining changes in lattice parameters, and hence strains, can be far more accurate. Troost *et al.* (1993) claim that by looking for changes in the angles between major zone axes, a strain sensitivity of about 0.1% is achievable. To study the elastic strain associated with epitaxial growth, the basic set-up used by Troost *et al.* (1993) on $Si_{1-x}Ge_x$ epitaxial films grown on Si substrates involved etching away, *ex situ*, the epitaxial layer in one section of the sample. This allowed an SEM-based EBSD pattern to be collected from an unstrained standard, and from the film of interest, without any changes in the beam configuration geometry. The image from the unstrained standard could thus be directly compared with the image from the epitaxial film, allowing the determination of a change in angles between major zone axes. A straightforward method for doing this is to subtract the image of the unstrained substrate from that of the film (see Fig. 2.7). It is important to note that because this technique looks at the interzonal angles, it will be sensitive only to strains that distort the shape of the unit cell, and not to strains caused by hydrostatic expansion or contraction. With this procedure, Troost *et al.* (1993) found a marked reduction in the interzonal angle between the [100] and [111] zone axes of $Si_{1-x}Ge_x$, which provided evidence of a tetragonal elongation of 2.5% for $x = 0.34$ and at 1.0% for $x = 0.16$ along the [100] direction of the unit cell. High-resolution X-ray diffraction measurements support these conclusions.

A further enhancement of strain sensitivity, down to about 0.02%, was obtained by Wilkinson (1996) and Wilkinson *et al.* (2009), also on $Si_{1-x}Ge_x$ epitaxial films grown on Si substrates. A similar etching procedure was used to provide a reference pattern for unstrained Si. Wilkinson (1996); Wilkinson *et al.* (2009) increased the sensitivity of the measurement primarily by increasing the sample-to-screen distance from 30 to 140 mm, thereby increasing the angular resolution of the detector. Using a more light-sensitive camera and computing the cross-correlation between multiple zone axes in both the strain-free reference images and the strained images further enhanced

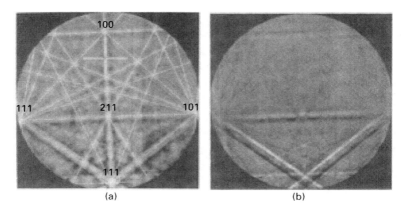

2.7 (a) Circular subsection of a BKP from Si after contrast enhancement by background subtraction with superimposed indexing of the lattice planes. (b) Image resulting from the subtraction of a BKP of pure Si (100) from a BKP of an epitaxial $Si_{0.66}Ge_{0.34}$ layer indicating the tetragonal elongation between the [001] and [111] zone axes of the strained layer. Reprinted with permission (Troost *et al.*, 1993).

the ability to distinguish small changes in the location of the zone axes. In addition, the use of multiple zone axes allowed the full strain tensor and lattice rotations to be calculated.

The determination of elastic lattice strain in growing films requires an image of the substrate prior to growth, from which any change in the interzonal angles can be detected as the film is growing. Therefore, the ability of a dual screen RHEED and Kikuchi capture allows the *in situ* study of elastic strain as a function of film thickness and other growth parameters, in real time.

2.7 Kinematic and dynamic scattering

Kikuchi diffraction patterns are the result of multiple elastic and inelastic scattering processes, and thus present an inherently dynamic scattering problem. However, kinematic diffraction from a point source within the sample can account for the geometry of the patterns, and is straightforward to simulate. Obtaining patterns that can be easily compared with experimental images requires a fairly large number of lattice planes to be included. It is the dynamical scattering that leads to interference terms that produce the more complicated features, particularly when two or more Kikuchi lines intersect. Generally the manifestations of dynamical scattering are found on the low-order Kikuchi bands, which show intensity irregularities along the band edges and high contrast within the bands.

The dynamical models for electron diffraction start with the Bloch wave approach to describe the diffraction of electrons by the crystal lattice. The lattice provides a three-dimensional diffraction grating for the electrons that are incident on the sample. Depending on the incidence angle of an incoming plane wave relative to the lattice planes, the diffracted electrons interact differently with the atoms which constitute the crystal. This is the basic description of SEM-based ECPs (Reimer, 1985). A parallel beam of electrons interacts with the sample and produces a given intensity of electrons that are backscattered. As the angle of the incident beam of electrons is changed, the backscattered intensity changes, and a pattern is built up that looks very similar to a BKP. BKPs are, in fact, considered to be the reciprocal of ECPs (Reimer, 1985). The reciprocity means that the observed intensity at a point detector inside the sample, after diffraction of an incoming plane wave, is the same as for a detected plane wave intensity after diffraction of waves emanating from a point source inside the sample.

Winkelmann *et al.* (2007) created a three-step model of the dynamic scattering in BKP. The first step is the diffraction of a focused incident beam, which produces a distribution of electrons within the sample. This diffracted electron distribution is then subject to an inelastic process, which is accounted for in the model by quasi-elastic backscattering from atomic nuclei. The final step is the reciprocal of the ECP model, where the quasi-elastic backscattering generates point sources of electrons within the sample that are then subject to diffraction from the surrounding lattice. The cross-section for the quasi-elastic backscattering from the atomic nuclei and the probability of finding an electron at the atomic nuclei provide the source strength for the outgoing diffraction. In addition, the quasi-elastic backscattering also destroys the coherence between the incident beam diffraction and the outgoing beam diffraction. Wilkelman *et al.*'s application of the dynamical theory provided simulated patterns which very closely reproduced experimental BKPs of GaN, including HOLZ ring effects and zone axis fine structure, and provides a framework for more quantitative analysis of BKPs (see Fig. 2.8). Recent additional work by Winkelman (2010) suggests that the depth-distribution of backscattered electrons should also be taken into account in quantitative descriptions of Kikuchi pattern formation.

Although dynamic scattering models for BKPs assume quasi-elastic scattering (Omoto *et al.*, 2002; Winkelmann *et al.*, 2007; Winkelmann, 2010), there is experimental evidence showing that electrons which lose up to 20% of the incident beam energy still participate in the formation of the Kikuchi pattern (Deal *et al.*, 2008). Went *et al.* (2009) recently showed spectroscopic evidence that in BKPs there is an elastic peak that is separated from the loss distribution. They also found that electrons with less than 2.5 eV of energy loss, expected for the case of thermal diffuse scattering and still within the elastic peak, provide the most contrast. In essence, the experimental pattern should be considered as a sum of different patterns,

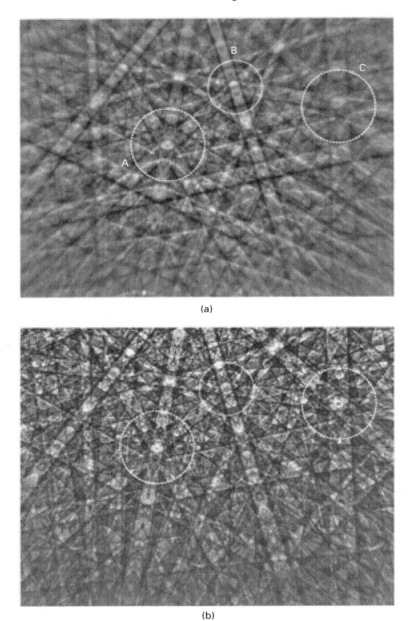

2.8 BKP of GaN 0001 at 20 kV: (a) experimental pattern; (b) dynamical simulation. In area B, attention is drawn to the HOLZ ring at approximately two-thirds of the radius of the area. The fine structure in areas A and C is strikingly similar in the two patterns. Reprinted with permission (Winkelmann, 2010).

each calculated for a specific energy loss, but strongly weighted towards quasi-elastic scattering.

As mentioned earlier, ECPs and BKPs are considered to be reciprocal measurements, and the general forms of the two patterns are indeed very similar. However, ECPs tend to show much more detail than the BKPs. This difference is a result of the much larger energy spread of electrons involved in the multi-step BKP pattern formation, as compared with the well-defined energy of the incident beam that generates the ECPs (Wilkinson and Hirsch, 1997).

As an alternative to a fully dynamic simulation, Zaefferer (2007) argues that the limited resolution of most cameras, and the different coherent scattering path lengths for all the electrons, lead to a blurring of the dynamical diffraction effects, resulting therefore in the appearance of bands of average intensities. Zaefferer showed that a significant improvement in the intensity calculations for BKPs over intensities that are proportional to the square of the structure factor, i.e. kinematic intensities, can be achieved by applying a relatively simple correction function to the kinematical intensities, originally proposed by Blackman (1939). This correction is appropriate only outside principal zone axes; nevertheless that accounts for a significant portion of the BKPs and therefore this approach may be a quick starting point for comparing experimental patterns to possible structures.

2.8 Epitaxial film structure determination

As discussed earlier, Wilkinson demonstrated that it is possible to obtain sensitivity to strains of approximately 0.02% on thick $Si_{1-x}Ge_x$ epitaxial layers grown on planar Si substrates (Wilkinson, 1996). The method used – analysis of interzonal angles – is limited to strains that distort the shape of the unit cell. However, unit cell distortion is of primary interest in understanding the tetragonal distortion that can occur with epitaxial growth; therefore interzonal angles are one method for tracking epitaxially driven changes to the structure of thin films.

Another approach to exploring possible symmetry changes caused by epitaxy is highlighted by Baba-Kishi's comparison of ZnS and $CuFeS_2$ (Baba-Kishi, 1992, 2002). ZnS is cubic and $CuFeS_2$ is tetragonal with a doubled ZnS-type structure. The symmetry features of BKPs in the cubic ZnS and those of the tetragonal $CuFeS_2$ are very similar, since the tetragonal distortion in $CuFeS_2$ is very small. However, it is possible to distinguish subtle changes to the BKP. The Kikuchi lines from $CuFeS_2$ that intersect within the Kikuchi bands generated from the (02$\underline{4}$) plane highlight the slight tetragonal distortion. These Kikuchi line intersections are slightly offset from the center of the Kikuchi bands (see Fig. 2.9) and indicate that for $CuFeS_2$ these planes are not mirror planes, as they would be for the cubic ZnS structure.

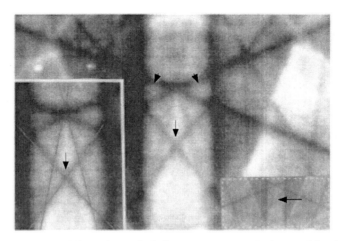

2.9 Kikuchi lines from CuFeS$_2$ intersecting within the Kikuchi band formed from the (0$\underline{2}$4) plane, showing a shift towards the right with respect to the large arrow which points to the center of the band. This shift at the intersection point originates from the tetragonal structure of CuFeS$_2$. Inset: lines drawn over the Kikuchi lines to help indicate the asymmetry of the intersections with respect to the large black arrow. Reprinted with permission (Baba-Kishi, 2002).

Changes in the atomic positions within a unit cell may not be readily detected by these analysis techniques. Therefore it may also be of interest to look at the differences in the intensity of specific Kikuchi bands as a route to explore other epitaxially influenced changes in the structure of thin films. To sort out what to look for in the experimental Kikuchi pattern in order to help distinguish possible changes in the crystal structure of a film due to effects of epitaxy, a simulation is needed. If changes to the patterns can be found away from principal zone axes, then the corrected kinematic simulation scheme proposed by Zaefferer (2007) (and discussed above) may be an appropriate starting point.

An example of a possible epitaxial influence on the structure of a grown film is the potential buckling of the Cu–O planes in the high-pressure phase of SrCuO$_2$ when grown with compressive strain. SrCuO$_2$ is the end member ($n = \infty$) of the homologous high-pressure series Sr$_{n-1}$Cu$_{n+1}$O$_{2n}$ (Hiroi *et al.*, 1991). Owing to the simplicity of the crystal structure and the planar nature of its Cu and O ordering, it is often called the parent compound of the high-temperature superconductors (Jorgensen *et al.*, 1993) and is considered to be a two-dimensional quantum magnet (Yasuda *et al.*, 2005). An extremely large change in the *c*-axis of this material was seen when it was grown by epitaxy with reduced *a* and *b* lattice parameters (Ingle *et al.*, 2002). One possible explanation for this anomalous expansion in the *c*-axis is the buckling of the Cu–O planes, suggested by Vailionis *et al.* (1997). The proposed buckling of the Cu–O planes occurs by the motion of just the oxygen atoms, and as such it is extremely hard to see in X-ray diffraction. However, Kikuchi diffraction may be an ideal tool to look at the potential of an epitaxially driven change in the dimensionality of the Cu–O planes.

Figure 2.10 shows several changes in the Kikuchi pattern that are expected between planar and buckled Cu–O planes in $SrCuO_2$, as simulated using the corrected kinematic scheme described by Zaefferer (2007). These Kikuchi patterns are centered around the ($\underline{1}30$) zone axis and indicate several possible changes in the pattern if the Cu–O planes were to buckle by the oxygen atoms moving in and out of the a–b plane by 10% of the c lattice constant (Fig. 2.10(b)) in contrast to a planar Cu–O structure (Fig. 2.10(a)). In this part of the Kikuchi pattern buckling causes an increase in intensity of the scattering from the {224} planes, indexed according to the enlarged unit cell required to describe buckling and highlighted in Fig. 2.10(b). Furthermore, there is also significant reduction in intensity expected for the (001) and (003) planes, as well as for the cubic-indexed {204} planes from the buckled structure (highlighted in Fig. 2.10(a)).

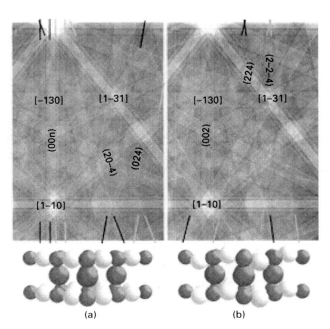

2.10 The calculated Kikuchi patterns, and structures, of the high-pressure phase of $SrCuO_2$ without buckled Cu—O planes (a), and with buckled Cu—O planes (b). The light and dark lines at the top and bottom of the image are guides to the eye of the respective excess and deficient lines that define the edges of the Kikuchi bands that change between the two images. Reprinted with permission (Ingle et al., 2010).

There are a number of other expected changes in the Kikuchi pattern for the buckling of the Cu–O plane in SrCuO, but these are primarily in the high-index planes. While high-index planes are present in the image, they not only have an inherently lower intensity but they are also overlaid with the more intense low-index planes, and locations in the pattern are not always present that allow them to be seen. However, these simulations suggest that even subtle changes in the structure of epitaxially grown films might be quite distinguishable in an *in situ* and real-time dual-screen RHEED/Kikuchi pattern collection set-up. In addition, the ability to collect a Kikuchi pattern from a well-characterized substrate just prior to the growth of a film provides a remarkable opportunity to minimize unknown experimental and geometrical effects in the patterns and to clarify the significant aspects of the simulations. This affords a tremendous advantage to the final analysis and interpretation of the Kikuchi patterns collected from an epitaxial film.

The use of more recent multi-beam dynamic models by Wilkelmann *et al.* (2007), which include improvements in the handling of inelastic scattering processes and the description of the excitation process, can provide much more detailed simulations. These models, along with continual improvements in hardware (Day, 2008) and software that are being applied to EBSD in SEMs, may allow a much more accurate and fine-grained approach to look for subtle changes in the atomic structure by inspecting and understanding the experimental Kikuchi patterns collected from thin epitaxial films.

2.9 Conclusion

The use of grazing incidence high-energy electrons during the growth of epitaxial films not only generates RHEED patterns from the elastically scattered electrons, but also generates Kikuchi patterns at larger scattering angles from the inelastically scattered electrons. These Kikuchi patterns can be analyzed to obtain an abundance of information about the crystallographic structure of the film. Of particular relevance to epitaxial film growth, the effects of strain and more generally the changes in space group and atomic positions of thin films can explored. These Kikuchi patterns can be collected in real-time, and simultaneously with the RHEED data with a dual-screen RHEED/Kikuchi collection set-up. The SEM-based EBSD community can provide a wealth of experience and has laid down a substantial amount of ground work that the film growth community can immediately use to help collect, analyze, and simulate Kikuchi patterns. In combination with standard RHEED analysis, the inelastically scattered electrons allow a remarkably complete set of data which help elucidate the full crystallographic structure of the potentially novel structures that can be formed by the influence of epitaxy on the growth of thin films.

2.10 References

M.N. Alam, M. Blackman, and D.W. Pashley. High-angle Kikuchi patterns. *Proceedings of the Royal Society of London, Series A*, **221**: 224, 1954.

K.Z. Baba-Kishi. A study of directly recorded RHEED and BKD patterns in the SEM. *Ultramicroscopy*, **34** (3): 205–218, 1990.

K.Z. Baba-Kishi. Electron-microscopy of the mineral chalcopyrite, $CuFeS_2$. *Journal of Applied Crystallography*, **25**: 737–743, 1992.

K.Z. Baba-Kishi. Measurement of crystal parameters on backscatter Kikuchi diffraction patterns. *Scanning*, **20** (2): 117–127, 1998.

K.Z. Baba-Kishi. Review – electron backscatter Kikuchi diffraction in the scanning electron microscope for crystallographic analysis. *Journal of Materials Science*, **37** (9): 1715–1746, 2002.

K.Z. Baba-Kishi and D.J. Dingley. Backscatter Kikuchi diffraction in the SEM for identification of crystallographic point groups. *Scanning*, **11** (6): 305–312, 1989a.

K.Z. Baba-Kishi and D.J. Dingley. Application of backscatter Kikuchi diffraction in the scanning electron-microscope to the study of NiS_2. *Journal of Applied Crystallography*, **22**: 189–200, 1989b.

S. Biggin and D.J. Dingley. General method for locating X-ray source point in kossel diffraction. *Journal of Applied Crystallography*, **10** (1): 376–385, 1977.

M. Blackman. On the intensities of electron diffraction rings. *Proceedings of the Royal Society of London Series A – Mathematical and Physical Sciences*, **173** (A952): 0068–0082, 1939.

H. Boersch. About bands in electron diffraction. *Physikalische Zeitschrift*, **38**: 1000–1004, 1937.

W. Braun. *Applied RHEED*. Springer, 1999.

W. Braun, L. Daweritz, and K.H. Ploog. Origin of electron diffraction oscillations during crystal growth. *Physical Review Letters*, **80** (22): 4935–4938, 1998a.

W. Braun, H. Moller, and Y.H. Zhang. Reflection high-energy electron diffraction during substrate rotation: a new dimension for *in situ* characterization. *Journal of Vacuum Science & Technology B*, **16** (3): 1507–1510, 1998b.

P.E. Champness. *Electron Diffraction in the Transmission Electron Microscope*. Microscopy Handbook Series no. 47. BIOS Scientific Publishers Ltd, 2001.

J.M. Cowley. *Diffraction Physics*. Elsevier Science BV, Netherlands, 1995.

A. P. Day. Spherical EBSD. *Journal of Microscopy – Oxford*, **230**: 472–486, 2008.

A. Deal, T. Hooghan, and A. Eades. Energy-filtered electron backscatter diffraction. *Ultramicroscopy*, **108**: 116–125, 2008. DOI 10.1016/j.ultramic.2007.03.010.

D. Dingley. Progressive steps in the development of electron backscatter diffraction and orientation imaging microscopy. *Journal of Microscopy – Oxford*, **213**: 214–224, 2004.

D.J. Dingley and V. Randle. Microtexture determination by electron back-scatter diffraction. *Journal of Materials Science*, **27** (17): 4545–4566, 1992.

D.J. Dingley and S.I. Wright. Determination of crystal phase from an electron backscatter diffraction pattern. *Journal of Applied Crystallography*, **42**: 234–241, 2009. DOI 10.1107/S0021889809001654.

D.J. Dingley, R. Mackenzie, and K.Z. Baba-Kishi. Applications of bkd for phase identification of crystals and strain measurement in materials. In P.E. Russel, editor, *Microbeam Analysis*, page 435. San Francisco Press, Inc., 1989.

D.J. Dingley, K.Z. Baba-Kishi, and V. Randle. *Atlas of Backscattering Kikuchi Diffraction Patterns*. Insitute of Physics, Bristol, 1995.

P.J. Dobson. An introduction to reflection high energy electron diffraction. In A. Howie and U. Valdre, editors, *Surface and Interface Characterization by Electron Optical Methods*, volume 191 of *NATO ASI Series B*, page 159, Plenum Press, 1988.

D.P. Field. Recent advances in the application of orientation imaging. *Ultramicroscopy*, **67**: 1, 1997.

M. Gajdardziska-Josifovska and J.M. Cowley. Brillouin zones and Kikuchi lines for crystals under electron channeling conditions. *Acta Crystallographica Section a*, **47**: 74–82, 1991.

T. Gareth. *Transmission Electron Microscopy of Materials*, Wiley, 1979.

C.J. Harland, P. Akhter, and J.A. Venables. Accurate microcrystallography at high spatial-resolution using electron backscattering patterns in a field-emission gun scanning electron-microscope. *Journal of Physics E–Scientific Instruments*, **14** (2): 175–182, 1981.

Z. Hiroi, M. Azuma, M. Takano, and Y. Bando. A new homologous series $Sr_{n-1}Cu_{n+1}O_{2n}$ found in the SrO-CuO system treated under high-pressure. *Journal of Solid State Chemistry*, **95** (1): 230, 1991.

Y. Horio. Kikuchi patterns observed by new astrodome RHEED. *e-Journal of Surface Science and Nanotechnology*, **4**: 118, 2006.

Y. Horio, Y. Hashimoto, and A. Ichimiya. New type of RHEED apparatus equipped with an energy filter. *Applied Surface Science*, **100**: 292–296, 1996.

A. Ichimiya and P.I. Cohen. *Reflection High Energy Electron Diffraction*. Cambridge University Press, 2004.

N.J.C. Ingle, R.H. Hammond, and M.R. Beasley. Molecular beam epitaxial growth of $SrCuO_2$: Metastable structures and the role of epitaxy. *Journal of Applied Physics*, **91** (10): 6371–6378, 2002. DOI 10.1063/1.1466876.

N.J.C. Ingle, A. Yuskauskas, A. Wicks, M. Paul, and S. Leung. Topical review: the structural analysis possibilities of reflection high energy electron diffraction. *Journal of Physics D – Applied Physics*, **43**: 133001, 2010.

J.D. Jorgensen, P.G. Radaelli, D.G. Hinks, J.L. Wagner, S. Kikkawa, G. Er, and F. Kanamaru. Structure of superconducting $Sr_{0.9}La_{0.1}CuO_2$ (T = 42K) from neutron powder diffraction. *Physical Review B*, **47** (21): 14654, 1993.

S. Kikuchi. Diffraction of cathode rays by mica. *Japanese Journal of Physics*, **5**: 83, 1928.

H.J. Kohl. In J.M. Sturgess, editor, *Proceedings of the 9th International Congress on Electron Microscopy (Toronto)*, page 198, 1978.

N.C.K. Lassen, D.J. Jensen, and K. Conradsen. Image-processing procedures for analysis of electron back scattering patterns. *Scanning Microscopy*, **6** (1): 115–121, 1992.

J.R. Michael and J.A. Eades. Use of reciprocal lattice layer spacing in electron backscatter diffraction pattern analysis. *Ultramicroscopy*, **81** (2): 67–81, 2000.

H. Nakahara, T. Hishida, and A. Ichimiya. Inelastic electron analysis in reflection high-energy electron diffraction condition. *Applied Surface Science*, **212**: 157–161, 2003. DOI 10.1016/S0169-4332(03)00057-6.

K. Omoto, K. Tsuda, and M. Tanaka. Simulations of Kikuchi patterns due to thermal diffuse scattering on MgO crystals. *Journal of Electron Microscopy*, **51** (1): 67–78, 2002.

V. Randle. *Microtexture Determination and its Application*. Bourne Press, 1992.

L. Reimer. *Scanning Electron Microscopy*. Springer-Verlag, 1985.

R.A. Schwarzer. Automated crystal lattice orientation mapping using a computer-controlled SEM. *Micron*, **28** (3): 249–265, 1997.

J.A. Small, J.R. Michael, and D.S. Bright. Improving the quality of electron backscatter diffraction (EBSD) patterns from nanoparticles. *Journal of Microscopy – Oxford*, **206**: 170–178, 2002.

P. Staib, W. Tappe, and J.P. Contour. Imaging energy analyzer for RHEED: energy filtered diffraction patterns and in situ electron energy loss spectroscopy. *Journal of Crystal Growth*, **201**: 45–49, 1999.

K.Z. Troost, P. Vandersluis, and D.J. Gravesteijn. Microscale elastic-strain determination by backscatter Kikuchi diffraction in the scanning electron-microscope. *Applied Physics Letters*, **62** (10): 1110–1112, 1993.

A. Vailionis, A. Brazdeikis, and A.S. Flodstrom. Observation of local oxygen displacements in CuO_2 planes induced by a misfit strain in heteroepitaxially grown infinite-layer-structure $Ca_{1-x}Sr_xCuO_2$ films. *Physical Review B*, **55** (10): R6152, 1997.

J.A. Venables and R. Binjaya. Accurate micro-crystallography using electron backscattering patterns. *Philosophical Magazine*, **35** (5): 1317–1332, 1977.

J.A. Venables and C.J. Harland. Electron backscattering patterns – new technique for obtaining crystallographic information in scanning electron-microscope. *Philosophical Magazine*, **27** (5): 1193–1200, 1973.

J.A. Venables, C.J. Harland, and Bin-Jaya. *Developments in Electron Microscopy and Analysis*, page 101. Academic Press, 1976.

R. von Meibom and E. Rupp. Wide-angle electron diffraction. *Zeitschrift für Physik*, **82**: 690, 1933.

Z.L. Wang. *Reflection Electron Microscopy and Spectroscopy for Surface Analysis*. Cambridge University Press, 1996.

O.C. Wells. Comparison of different models for the generation of electron backscattering patterns in the scanning electron microscope. *Scanning*, **21** (6): 368–371, 1999.

M. R. Went, A. Winkelmann, and M. Vos. Quantitative measurements of Kikuchi bands in diffraction patterns of backscattered electrons using an electrostatic analyzer. *Ultramicroscopy*, **109** (10): 1211–1216, 2009. DOI 10.1016/j.ultramic.2009.05.004.

A.J. Wilkinson. Measurement of elastic strains and small lattice rotations using electron back scatter diffraction. *Ultramicroscopy*, **62** (4): 237–247, 1996.

A.J. Wilkinson and P.B. Hirsch. Electron diffraction based techniques in scanning electron microscopy of bulk materials. *Micron*, **28** (4): 279–308, 1997.

A.J. Wilkinson, G. Meaden, and D.J. Dingley. Mapping strains at the nanoscale using electron back scatter diffraction. *Superlattices and Microstructures*, **45** (4–5): 285–294, 2009. DOI 10.1016/j.spmi.2008.10.046.

A. Winkelmann. Principles of depth-resolved Kikuchi pattern simulation for electron backscatter diffraction. *Journal of Microscopy – Oxford*, **239** (1): 32–45, 2010. DOI 10.1111/j.1365-2818.2009.03353.x.

A. Winkelmann, C. Trager-Cowan, F. Sweeney, A. P. Day, and P. Parbrook. Many-beam dynamical simulation of electron backscatter diffraction patterns. *Ultramicroscopy*, **107** (4–5): 414, 2007. DOI 10.1016/j.ultramic.2006.10.006.

C. Yasuda, S. Todo, K. Hukushima, F. Alet, M. Keller, M. Troyer, and H. Takayama. Neel temperature of quasi-low-dimensional Heisenberg antiferromagnets. *Physical Review Letters*, **94** (21): 217201, 3 2005. 217201.

S. Zaefferer. On the formation mechanisms, spatial resolution and intensity of backscatter Kikuchi patterns. *Ultramicroscopy*, **107** (2–3): 254, 2007. DOI 10.1016/j.ultramic.2006.08.007.

Part II
Photoemission techniques for studying thin film growth *in situ*

3
Ultraviolet photoemission spectroscopy (UPS) for *in situ* characterization of thin film growth

K. M. SHEN, Cornell University, USA

Abstract: This chapter discusses the application of ultraviolet photoemission spectroscopy (UPS) and angle-resolved photoemission spectroscopy (ARPES) to study the electronic structure of thin films synthesized *in situ* using film growth techniques such as molecular beam epitaxy (MBE) and pulsed laser deposition (PLD). The chapter first describes the principles of photoemission spectroscopy and the current experimental state of this technique. Then, particular examples such as UPS and ARPES studies of *in situ* thin films of $SrRuO_3$ and $La_{1-x}Sr_xMnO_3$ are detailed.

Key words: photoemission spectroscopy, angle-resolved photoemission spectroscopy (ARPES), correlated electronic materials, *in situ* thin film spectroscopy.

3.1 Introduction

One of the prime motivators for the growth of materials in thin film form is the potential for manipulating their electronic and magnetic properties. One of the forefronts of modern condensed matter physics is the search for new states of quantum matter with exotic physical properties, and thin films and superlattices present a unique opportunity for designing and realizing such novel materials. One way of achieving this control is to create atomically perfect interfaces between two different materials. Modern technologies such as the transistor or laser are dependent on interfacial phenomena. Remarkable scientific discoveries such as the integer and fractional quantum Hall effects (Nobel Prizes 1985, 1998) were discovered at semiconductor interfaces engineered using molecular beam epitaxy (MBE). Since 2005, interfacial phenomena at the atomically abrupt interface between dissimilar materials have yielded a flurry of breakthroughs. To fully benefit from the tailoring of these novel thin film interfaces and heterostructures, one needs to be able to characterize the state of electronic matter created in each system. In order to do so, it will be desirable to utilize sophisticated 21st century tools to achieve this aim, such as high-resolution angle-resolved photoemission spectroscopy (ARPES). The ARPES technique follows from the development of electron spectroscopy for chemical analysis (ESCA) pioneered by Kai Siegbahn (Nobel

Prize 1981) by probing not only the energy but also momentum distribution of the photoelectrons to directly observe band structure and Fermi surface of materials. Without the detailed information about the electronic structure that ultraviolet photoemission spectroscopy (UPS) and ARPES can provide, a deep understanding of how and why each thin film material exhibits the properties it does is lacking. In turn this greatly diminishes the capabilities for rational design of new electronic materials in thin film form.

By extracting electrons from within a material and measuring their momentum and energy states, ARPES can measure a quantity known as the electrons' 'Green's function', a quantity which encodes all the information about the collective motion of electrons, how they are entangled, and provides full knowledge of the electronic structure. Despite its power, ARPES has one major limitation: the requirement of atomically pristine surfaces which demands large, bulk single crystals which yield flat, well-defined surfaces when cleaved under vacuum. This is where photoemission spectroscopy on thin films grown *in situ* can provide an entirely new avenue into research into the electronic properties of thin films. Growth techniques such as MBE have the distinct advantage of occuring at very low partial pressures (10^{-7} or 10^{-8} torr) and yielding atomically flat surfaces. These points are key to achieving the atomically pristine, flat well-defined surfaces that are essential for ARPES.

One important aspect of correlated electronic systems, of which transition metal oxides are a major constituent, is that the electrons are strongly entangled, and it is their intricate collective motion that gives rise to their remarkable attributes. This is in contrast to conventional materials where the individual electrons essentially move as independent particles and thus cannot produce these effects. However, understanding this entanglement which is so essential to these properties is an extremely difficult scientific problem because our theoretical physical understanding is based largely on the non-interacting paradigm of materials such as silicon or copper.

3.2 Principles of ultraviolet photoemission spectroscopy (UPS)

Photoemission spectroscopy is a technique which has rather illustrious roots. The explanation of the photoelectric effect by Albert Einstein in 1905 has been widely heralded as one of the greatest scientific breakthroughs in human history. While his work on relativity captured more of the spotlight and the public's imagination, it is 'especially for his discovery of the law of the photoelectric effect' for which he was awarded the 1921 Nobel Prize. Moreover, the elucidation of the photoelectric effect opened the door to the quantum world, which has formed our basis for understanding all of modern science. It was not until the late 1950s and early 1960s that it was observed that the photoelectric effect could yield interesting information on the nature of the electronic states of the illuminated cathode material. Owing to the conservation of energy, it was noticed that the energy distribution of photoemitted electrons could provide information on the density of electronic states in the cathode material. In particular, the photoemission work of Berglund and Spicer (1964) on Cu and Ag showed the edges of the d bands at 2 and 4 eV below the Fermi energy, in agreement with the predictions of the non-interacting band theory. Later, as previously mentioned Kai Siegbahn would share the 1981 Nobel Prize in Physics for his development of electron spectroscopy.

The basic principles of UPS using photons in the energy range between 10 and 100 eV are the same as those of X-ray photoelectron spectroscopy (XPS). However, XPS is primarily used to study atomic core levels which are typically hundreds or even thousands of electron volts below the Fermi energy (E_F) which can provide crucial information about chemical composition, oxidation states, and chemical bonding. UPS is mainly employed to study the states near E_F (within 0–20 eV). The photoelectron cross-sections at these lower photon energies is much more favorable to states near E_F than X-rays and much higher energy resolutions ($\Delta E \sim$ meV) are achievable in UPS than in XPS, which is critical for studying low-lying states near E_F.

3.2.1 ARPES

As early as 1964, Kane argued that the momentum-dependent band structure could be mapped from the angle and energy dependence of the photoemission spectra. However, during the formative years of photoemission, all experiments were exclusively angle-integrated. In the early 1970s, work began in earnest on measuring the angular distribution of the photoelectrons. It was not until 1974 that the first angular dependent band-mapping was performed using photoemission by Traum, Smith, and DiSalvo (1974), which also used two-dimensional (2D) layered compounds TaS_2 and $TaSe_2$ to avoid the complications of the uncertainty in the transverse momentum, k_z, and were able to show good agreement with band structure calculations.

The energy resolution of photoemission experiments in the late 1970s and 1980s was in the order of 100 meV. In contrast, the thermodynamic properties of solids are determined by the electronic states within a thin strip of energy about $k_B T$ wide where k_B is Boltzmann's constant, around the Fermi energy. At room temperature, this corresponds to a thermal energy of 30 meV, but at low temperatures where phenomena such as superconductivity occur ($\sim$ 10 K), this corresponds to an energy of roughly 1 meV. Therefore, an energy resolution of 100 meV (corresponding to roughly 1000 K, in the order of the melting temperature of most solids) was clearly insufficient to address anything beyond the gross electronic structure of solids. However, substantial

58   *In situ* characterization of thin film growth

advances in detector technology in the late 1990s and early 2000s resulted in order-of-magnitude improvements in the energy and angular resolutions, along with the data acquisition efficiency. At present, multichannel angular and energy detection in normal ARPES experiments typically occur using $\Delta E \sim 5$ meV, $\Delta \theta \sim 0.2°$, and with over 100 angular channels acquired simultaneously. Using the latest generation of analyzers and laser-based light sources, sub-meV energy resolutions have recently been demonstrated (Kiss *et al.* 2008).

Figure 3.1 shows the energetics of the photoemission process in a simplified density-of-states picture. Within the solid, there is an electronic density of states governed by the intrinsic band structure and interaction effects, where states are filled up to the Fermi energy. There is a finite potential energy barrier between the first occupied electronic state (at $T = 0$) at E_F, and the vacuum level (the potential energy zero infinitely far away from the sample), which is the work function of the material, Φ, and this is what holds the

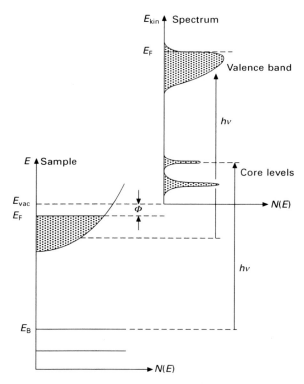

3.1 The left side of the figure shows energy levels in the crystal in the initial state. The photoemission process occurs with the absorption of a photon with energy $h\nu$. The right side of the figure shows the measured photoemission spectrum, starting from the vacuum level. From Damascelli *et al.* (2003).

electrons inside the crystal. Therefore, when an electron absorbs a photon (and does not experience any inelastic losses), the binding energy of the electron in the initial state relative to the Fermi energy, E_B, can be related to the measured kinetic energy, E_{kin}, by $h\nu - \Phi - E_F$. The work function in question is that of the detector, since all materials have slightly different work functions depending on the characteristics of the surface. If the work function of the sample and detector are different, then a potential will be set up between the sample and detector.

To relate the relationship of the photoelectron momentum to the momentum of the photohole left in the crystal, we consider an electron in a crystal with a well-defined quasimomentum K, and energy E_B. If the incident photon has enough energy to promote the electron above E_{vac}, then the electron can be photoemitted into free space. Because the electron has to propagate a macroscopic distance (~ 1 m) to the detector, it can essentially be treated as a free plane wave with $E = \hbar k^2/2m$. Therefore, by knowing the takeoff angles of the electron relative to the crystal axes and the detector, we can determine the momentum wavevector of the outcoming photoelectron. The geometrical considerations for the photoemission process are illustrated in Fig. 3.2. From this we can use momentum conservation to relate this to the quasimomentum of the electron in the initial state. We can roughly generalize

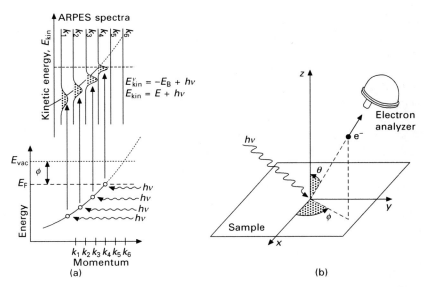

3.2 (a) Illustration of optical (direct) transitions from electrons in a band in the crystal, to photoelectrons in free space. (b) ARPES and measured observables, being photoelectron counts, photoelectron kinetic energy, and polar angles of photoelectron relative to sample crystal axes. From Damascelli et al. (2003).

the real situation to a semi-infinite crystal (infinite in the lateral directions, but with a discontinuous step – the crystal surface – in the z direction). In this case, the momentum in the lateral direction, $k_{parallel}$ is conserved due to translational symmetry, but the momentum in the z direction is clearly not conserved, owing to the work function step potential at the surface (the 'lost' k_z momentum of the photoelectron across the barrier is then taken up by a recoil of the crystal). Therefore, we can obtain an exact measure of the in-plane wavevectors of the electron in the initial state, $k_{parallel}$, but extracting k_z is more challenging. To determine k_z, one should determine the so-called 'inner potential' of the crystal, which is essentially determining which photoelectron kinetic energy corresponds to $k_z = 0$ (the (0,0,0) point) (Hufner 1995). This is typically rather involved and requires spanning a wide range of incident photon energies, $h\nu$, in real measurements.

It is generally most appropriate to view the experimentally measured photocurrent as a photoinduced transition between electron initial and final states. In UPS, the incoming photon carries negligible momentum (for $h\nu = 10\,eV$, $Q \sim 5 \times 10^{-3} Å^{-1}$), so all transitions are essentially direct, but this should clearly not be the case for XPS. To reach an available final state, the photoelectron should then be translated by a reciprocal lattice vector, G, so that the wavevector of the outgoing photoelectron $K = k + G$, demonstrating that the photoemission process requires the presence of a lattice potential to conserve momentum. The above analysis allows us to relate the photoelectrons measured at some kinetic energy and momentum to electrons with some different binding energy and momentum in the initial state, inside the crystal before the photoemission process. However, it does not tell us the direct relationship between the electronic states of the N and $N-1$ system. For this, one must make further assumptions beyond simple kinematics and conservation relations to obtain meaningful information about the $N-1$ system of the photohole from the experimentally measured photocurrent.

Rigorously speaking, the best theoretical description of the photoemission process is the so-called 'one-step model'. In the one-step model, the photoemission process is treated as a single, entirely quantum-mechanical process. However, the phenomenological model which is nearly exclusively used in the experimental photoemission community is the 'three-step' model, popularized by Berglund and Spicer in 1964. The steps in the three-step model are as follows:

1. Optical excitation of the electron inside the solid.
2. Transport of the photoelectron to the surface.
3. Escape of the photoelectron into the vacuum.

The independent steps in the three-step model are shown in Fig. 3.3. In this semiclassical picture, only the first step is treated quantum mechanically, and the three steps are treated as independent of one another, in contrast

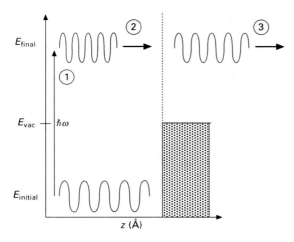

3.3 Illustration of the three-step model, with (1) optical excitation of an electron in the bulk, (2) transport of the photoelectron to the surface, and (3) escape of the photoelectron into vacuum free space.

to the correct quantum-mechanical one-step approach. The beauty of the three-step model is that it actually removes the complicated photoemission process itself from photoemission. The three-step model effectively reduces the photoemission process to an optical excitation. Steps (2) and (3) in the three-step model are simply attenuation factors and kinematic considerations, respectively, leaving (1) as the only nontrivial process. Given the dramatic simplifications of the three-step model, it works remarkably well and provides surprisingly accurate results.

Here we will discuss a few additional aspects of the photoemission process. First, photoemission is a highly surface-sensitive process because of the strong Coulomb interactions between the photoelectron and the rest of the electrons left in the solid, with a typical mean free path on the order of 1 nm for photon energies in the vacuum ultraviolet (VUV). Because the mean free path depends on the cross-section of the photoelectrons to scatter with plasmons, the mean free path should be even lower for 'bad metals' with low plasmon energies such as many strongly correlated transition metal oxides. This high degree of surface sensitivity is the reason for extremely stringent ultra-high vacuum conditions necessary for reliable, high-resolution ARPES experiments, and experiments are typically performed at pressures of better than 5×10^{-11} torr. Samples are usually cleaved *in situ* at low temperatures and base pressure to ensure an atomically clean surface layer. Even at these pressures, we can occasionally observe surface degradation, although this varies significantly on the particular compound and contaminants in the vacuum chamber.

3.2.2 Instrumentation and light sources for high-resolution photoemission spectroscopy

ARPES experienced a renaissance in the late 1990s and early 2000s with the advent of 2D multiplexing analyzers by the Swedish company VG Scienta which could measure a range of electron kinetic energies and angles simultaneously. The way that traditional electron analyzers worked was that their angular acquisition was based on a small pinhole aperture. The position of the pinhole relative to the sample normal determined the photoelectron K, and the solid angle subtended by the pinhole determined Δk. The problem with this approach is that all electrons outside this pinhole are discarded, and only one wavevector can be measured at a time. The great advantage of the Scienta analyzers is that the electron lens system for these analyzers allowed for parallel detection of many (~100) angular channels, thereby increasing the measurement efficiency by about two orders of magnitude. Today, such multiplexing analyzers are quite common and easily achieve energy resolutions of $\Delta E = 1$ meV and simultaneously measure angular ranges greater than 30 degrees in parallel. The electrons pass through the entrance aperature, and then through a three-element lens system, through the entrance slits, and then the two concentric hemispheres housed inside the large stainless steel dome (all electron lens elements are electrostatic). The electrons are finally detected at the bottom of the analyzer, after the electron signals are amplified by a pair of microchannel plates (MCPs) and a phosphor screen. Flashes on the phosphor screen are then detected *ex situ* by a CCD camera looking through the window, visible at the bottom flange. An illustration of a Scienta analyzer operating at a synchrotron beamline is shown in Fig. 3.4.

A major barrier to performing UPS is the relative dearth of appropriate UV and VUV photon sources. As a result, many UPS and ARPES systems are

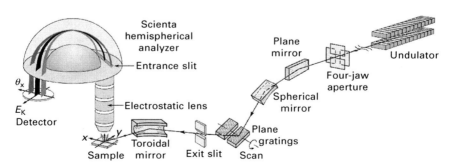

3.4 Schematic of Scienta analyzer on a beamline. Photoelectrons with different take-off angles are imaged onto the long entrance slit, and pass through the concentric hemispheres, and strike the microchannel plates. From Damascelli *et al.* (2003).

located at synchrotrons where at undulator insertion devices, tunable photon energies can be achieved with high photon fluxes ($\sim 10^{12}$ photons/s mm^2) within a narrow ($\sim$ meV) energy bandwidth. Synchrotron sources remain the light source of choice for ARPES, as no other sources can match their flexibility in terms of photon energies and polarizations. However, because of the scarcity and relatively small number of synchrotrons, laboratory-based UV and VUV sources present a highly desirable alternative. These sources must exceed the energy of the work function (typically 4–5 eV), and such light sources are relatively scarce. The most common light sources for laboratory-based UPS are plasma discharge (He, Ar, Xe) lamps which have discrete emission lines with high brightness ($\sim 10^{12}$/s mm^2) and narrow spectral widths (~ 1 meV). The main disadvantage to plasma discharge lamps is the lack of tunability in photon energies which can be essential for many experiments. More recently, VUV lasers which use higher harmonic generation (HHG) have come to the forefront. Although currently restricted in photon energy tunability and limited to fairly low photon energies (5–7 eV), they have the distinct advantage of superior spectral linewidths and brightness over plasma discharge lamps. In addition, laser sources can be utilized for time-resolved experiments such as pump-probe measurements. However, at present, the wavelengths available to VUV lasers is rather restricted, although this is likely to change as the technology matures.

3.3 Applications of UPS to thin film systems

3.3.1 SrRuO$_3$

SrRuO$_3$ has drawn wide interest for thin film studies and applications for a number of major reasons. First, SrRuO$_3$ is a ferromagnetic metal (T_c = 160 K) which is unusual for a 4d transition metal oxide, but also exhibits unconventional 'bad metal' behavior at elevated temperatures. The Ruddlesden–Popper series of layered ruthenates Sr$_{n+1}$Ru$_n$O$_{3n+1}$, of which SrRuO$_3$ is the three-dimensional end-member, display a remarkable evolution in magnetic and electronic properties as a function of dimensionality, n. The single layer n = 1 compound, Sr$_2$RuO$_4$, is an exotic spin triplet superconductor with a time-reversal symmetry breaking ground state (Mackenzie and Maeno 2003). The n = 2 bilayer compound, Sr$_3$Ru$_2$O$_7$ exhibits quantum critical metamagnetism and an apparent electronic nematicity where the magnetoresistance breaks $C4$ symmetry (Borzi *et al.* 2007). Ferromagnetic tendencies are enhanced by increasing the number n of RuO$_2$ sheets per unit cell. In addition, Ca substitution for Sr dramatically alters the magnetic and electronic properties of all known members of the Sr$_{n+1}$Ru$_n$O$_{3n+1}$ ruthenates, despite the fact that Ca and Sr are isovalent (for instance, CaRuO$_3$ is a paramagnetic metal, and Ca$_2$RuO$_4$ is an antiferromagnetic insulator). In addition, spin–orbit interactions and 4*d* orbital degeneracies are known to be important factors. Therefore, $SrRuO_3$ and other members of the ruthenate family present a wealth of exotic electronic and magnetic behaviors which are ripe for investigation. From the thin film standpoint, $SrRuO_3$ is also an excellent material for investigation since epitaxial thin films of very high quality can be synthesized. Until very recently, many properties of films (such as resistivity) were found to exceed even those of the best bulk single crystals. In addition, the low resistivity of $SrRuO_3$ films make it an excellent candidate material for thin film applications such as electrode materials or magnetic tunnel junctions. In addition, the pseudocubic perovskite structure makes $SrRuO_3$ suitably lattice matched to other common transition metal oxides such as $SrTiO_3$, a common substrate material for $SrRuO_3$.

Because the electronic properties of $SrRuO_3$ are still not well understood, UPS and ARPES studies of $SrRuO_3$ are essential for answering important basic questions about the electronic structure, such as whether first-principles calculations can capture an accurate description of this complex correlated material which exhibits ferromagnetism and three-fold orbital degeneracy, or the nature of the large effective masses and the strength of electron–electron correlations. However, because of its pseudocubic structure, cleaving single crystals does not yield the flat, well-defined surfaces necessary for ARPES. Therefore, measurements of atomically flat thin film samples synthesized *in situ* are essential for performing ARPES measurements of the quasiparticle band structure and Fermi surface. Moreover, because of the surface sensitivity of the ARPES technique (inelastic mean free path ~ 1 nm), these thin film samples cannot be removed from ultra-high vacuum (UHV) conditions before measurement; exposure of an atomically pristine surface to atmosphere can render a surface unmeasurable within nanoseconds. Therefore, the synthesis, transfer, and measurement stages must all take place under strict UHV conditions.

The earliest UPS measurements of $SrRuO_3$ were performed by Fujioka *et al.* (1997) and Okamoto *et al.* (1999) on polycrystalline samples of $SrRuO_3$ which were sintered as pellets and scraped *in situ*. Based on comparisons to the density of states obtained from band structure calculations, Okamoto *et al.* were able to obtain a mass enhancement factor of approximately m^*/m_b ~ 4.4, which appeared consistent with estimates from transport measurements. On the other hand, based on this analysis, it is difficult to determine the origin of this fairly large mass enhancement, although it has been speculated by Fujioka *et al.* that this may be due to electron–electron interactions. On the other hand, Mazin and Singh (1997) have argued that electron–magnon and phonon couplings in $SrRuO_3$ could be anomalously large due to the important role of oxygen in forming the ferromagnetic state. A major limitation of performing photoemission spectroscopy on polycrystalline samples is the lack of any momentum-resolved information. Therefore, developing a method for

Ultraviolet photoemission spectroscopy (UPS)

pursuing *in situ* photoemission studies of $SrRuO_3$ thin films will be critical for in-depth studies of the electronic structure of $SrRuO_3$.

Using a novel combination of both PLD and MBE, Siemons *et al.* (2007) were able to synthesize a variety of $SrRuO_3$ thin films with varying degrees of Ru stoichiometry and quantified the strong dependence of the Ru stoichiometry (deficiency) on the unit cell volume and c-axis lattice constant as well as with the residual resistivity ratio (RRR), as shown in Fig. 3.5. They

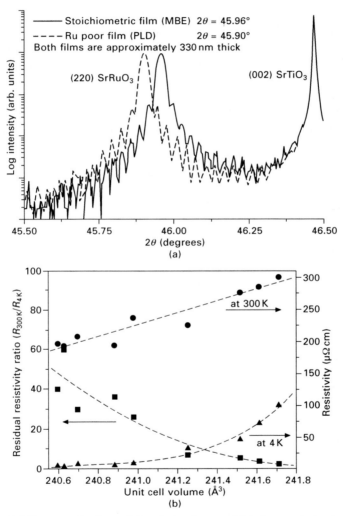

3.5 Characterization of the dependence of (a) the c-axis lattice constant and (b) the resistivity and residual resistivity ratio (R_{300K}/R_{4K}) on the Ru stoichiometry of $SrRuO_3$ thin films grown by PLD and MBE, as characterized by Siemons *et al.* (2007).

determine that films grown by PLD are typically ruthenium deficient with poor (low) residual resisitivities and can have an expaned unit cell volume, and could synthesize thin films with RRRs as low as 5 or less (PLD, high oxygen pressures) or up to 40 (MBE, low oxygen pressures). It is interesting to note that the Curie temperature is only very weakly dependent on Ru stoichiometry, and therefore should not be used as a metric for determining the quality of SrRuO$_3$ thin films. In addition, using *in situ* photoemission spectroscopy, Siemons *et al.* clearly demonstrate that the quality of the UPS spectra of these thin films near E_F ($E < 1$ eV) is strongly dependent on the Ru stoichiometry. In particular, the spectral intensity near E_F (constituted primarily of the Ru d_{xy}, d_{yz}, and d_{xz} orbitals) is strongly suppressed with respect to the background at higher binding energies as shown in Fig. 3.6. Therefore, measurements which attempt to extract the quasiparticle mass enhancement based solely on comparisons of spectral weight intensities and comparisons with theoretical densities of states must necessarily take into account the potentially sizable effects of Ru non-stoichiometry on the photoemission spectra.

As mentioned earlier, the electronic and magnetic properties of the Ruddlesden–Popper series ruthenates are known to depend critically on dimensionality. In response to this, ultrathin films of SrRuO$_3$ grown on SrTiO$_3$ have been investigated and an evolution in their electronic and

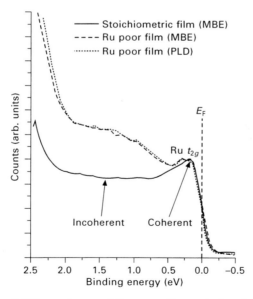

3.6 Dependence of the near E_F spectral weight on the Ru stoichiometry in thin films of SrRuO$_3$ as measured by *in situ* UPS using a He I discharge lamp. From Siemons *et al.* (2007).

magnetic properties have been reported by increasing the SrRuO$_3$ film thickness within the first 10 monolayers (MLs) of growth. Xia *et al.* (2009) have determined through a combination of polar Kerr effect measurements and resistivity that below a SrRuO$_3$ film thickness of 4 MLs, the material's ferromagnetic properties appear to be heavily suppressed, suggesting the formation of antiferromagnetism, while above 4 MLs, bulk-like behavior is reported. In conjunction with this ferromagnetic suppression, films thinner than 4 MLs exhibited insulating behavior in resistivity measurements, shown in Fig. 3.7. These findings are consistent with earlier reports by Toyota *et al.* (2005) which studied the evolution of the photoemission spectra of SrRuO$_3$ films as a function of thickness. As shown in Fig. 3.8, below approximately 6 MLs, there is little to no spectral intensity near E_F, while above 10 MLs the spectra are similar to much thicker (100 ML) films. This again points to some kind of metal–insulator transition around 4 MLs which also appears to be coincident with a dramatic change in the ferromagnetic properties of the material.

The work by Siemons and Toyota clearly establishes the importance of *in*

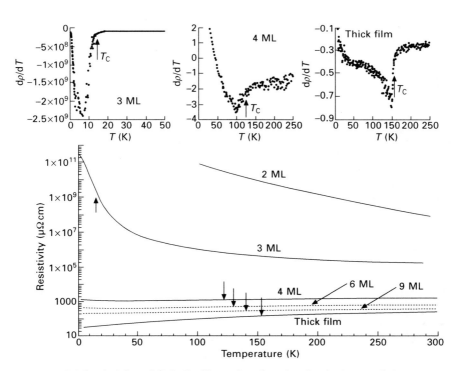

3.7 Resistivity of SrRuO$_3$ films showing the dependence of the resistivity and the ferromagnetic transition on the film thickness in monolayer (ML). From Xia *et al.* (2009).

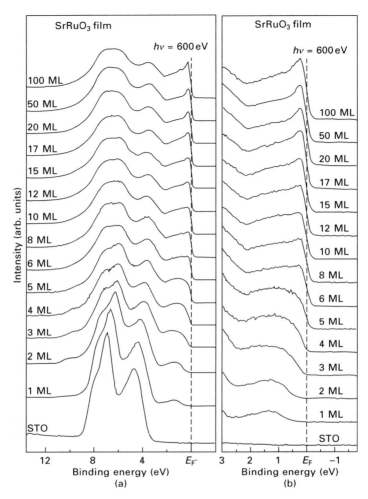

3.8 In situ photoemission valence band spectra of $SrRuO_3$ grown by PLD as a function of nominal film thickness, showing depleted spectral weight near E_F below 5 ML. From Toyota *et al.* (2005).

situ photoemission spectroscopy on uncovering the complex electronic structure of $SrRuO_3$. However, there is still much to be learned about the underlying electronic structure which may come from future angle-resolved measurements as well as additional thickness dependence studies of ultrathin $SrRuO_3$ films. At present, angle-resolved photoemission measurements of $SrRuO_3$ have not yet been conducted, although these recent successes of *in situ* angle-integrated photoemission measurements on $SrRuO_3$ and *in situ* ARPES on other thin films such as $La_{1-x}Sr_xMnO_3$ suggest that future ARPES measurements on $SrRuO_3$ films are not far off on the horizon. ARPES measurements will be key in answering critical questions about the intrinsic electronic structure of $SrRuO_3$, such as whether density functional theory can adequately predict the electronic structure of such a complex, correlated multi-band ferromagnet, or what is the true nature (electron–electron or electron–boson coupling) of the large mass renormalization observed in $SrRuO_3$. However, a major consideration will be that accounting for Ru stoichiometry will be particularly critical for high-quality ARPES measurements. A high concentration of Ru vacancies will introduce a large amount of scattering which would broaden and mix otherwise well-defined eigenstates in momentum space, meaning the materials requirements for ARPES will be even more stringent than for angle-integrated UPS.

3.3.2 $La_{1-x}Sr_xMnO_3$

The $Re_{1-x}Ae_xMnO_3$ manganites (with Re a rare earth, and Ae an alkaline earth metal) have been at the forefront of the field of strongly correlated electrons due to their host of electronic phases and dramatic phase transitions. Most well known amongst these properties is colossal magnetoresistance (CMR), where the electrical resistance can drop by more than six orders of magnitude with the application of a magnetic field. $La_{1-x}Sr_xMnO_3$ (LSMO) shows excellent metallic behavior and the highest Curie temperature of all $Re_{1-x}Ae_xMnO_3$ variants, and is predicted theoretically to be 100% spin polarized. Based on these properties, LSMO is a potentially ideal material for the development of new spintronics devices, particularly magnetic memory and magnetic sensors. Toward this end, CMR effects in LSMO tunnel junctions and the integration of semiconductors into LSMO-based devices have been demonstrated, and a great deal of interest and work has been devoted to synthesizing epitaxial thin films of LSMO and other manganite compounds. Recently, there has also been great activity in using LSMO as a building block for novel heterostructured materials. Cation-ordered versions of LSMO, where layers of $LaMnO_3$ are alternately stacked with layers of $SrMnO_3$, have shown new and dramatic metal–insulator transitions, increased ordering temperatures, and may generate spin-polarized two-dimensional electron gases at the $LaMnO_3$–$SrMnO_3$ interfaces.

Due to its three-dimensional pseudocubic structure, ARPES measurements of bulk single crystals of LSMO remain inaccessible because of the inability to produce a well-defined crystallographic surface upon cleaving. Therefore, like the case of $SrRuO_3$, atomically flat thin film samples of LSMO grown *in situ* using MBE or PLD provide the only route to accessing the electronic structure of these materials. The importance of understanding the complex electronic structure of this strongly correlated material has led to various research groups performing *in situ* ARPES studies of thin films of LSMO.

The first *in situ* ARPES studies of thin films of LSMO were reported by Shi *et al.* (2004) and later by Chikamatsu *et al.* (2006). Thin films were synthesized using pulsed laser deposition on $SrTiO_3$ substrates. Measurements were conducted using an energy resolution of 150 meV and an angular resolution of 0.5° for the work of Chikamatsu *et al.* and 40 meV and 0.2° for Shi *et al.* Chikamatsu *et al.* demonstrated for $x = 0.4$ that the valence band (within ~ 10 eV of the Fermi level) exhibits generally good agreement with band structure calculations based on a local density approximation plus on-site Coulomb interactions (LDA + U), shown in Fig. 3.9. Close to E_F (within 1 eV), only broader features with weak spectral weight could be observed near E_F. While the approximate dispersion of these features qualitatively tracked some of the LDA + U predictions, important questions persist about the nature of these near E_F states. First, the reason for the dramatically reduced near E_F spectral weight is still unclear, although it has been suggested that this spectral weight suppression could be the signature of polaronic behavior due to strong electron–phonon interactions, shown in Fig. 3.10. However, it may be conceptually challenging to reconcile small

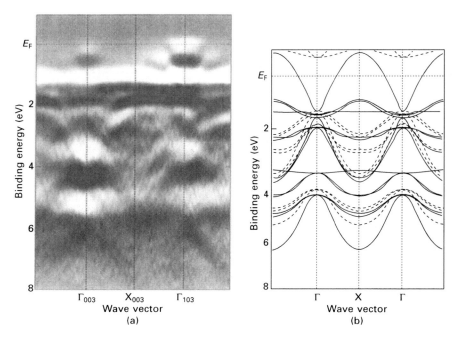

3.9 In situ ARPES measurements of the valence band of $La_{1.6}Sr_{0.4}MnO_3$ thin films, showing the (a) second-derivative plots of the valence band and (b) band structure calculations using the local density approximation plus on-site Coulomb interactions (LDA + U). Solid and dashed lines correspond to spin majority and minority bands respectively. From Chikamatsu *et al.* (2006).

Ultraviolet photoemission spectroscopy (UPS) 71

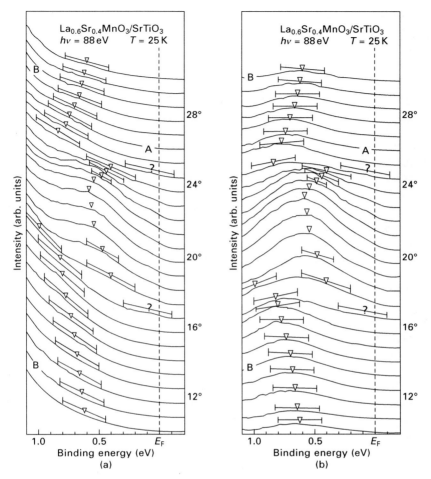

3.10 ARPES measurements of the near E_F spectra of $La_{1.6}Sr_{0.4}MnO_3$ thin films enlarged around the G ($k = (0,0,0)$) point. The open symbols represent the approximate energy positions of possible dispersive features. Panel (a) shows raw data; panel (b) shows data after subtraction of a high binding energy background. 'A' denotes the dispersive low energy band, while 'B' denotes the weakly dispersive high energy feature. From Chikamatsu et al. (2006).

polaron formation (and subsequently very large effective masses) with its reasonably metallic behavior at low temperatures ($\rho \sim 10^{-5} \Omega$ cm).

The work by Shi et al. on films of $x = 0.33$ grown by PLD also revealed that the large-scale electronic structure in the valence band also agreed reasonably well with predictions, but similar to the work by Chikamatsu et al., the near E_F electronic structure remained somewhat surprising. In particular, in both Chikamatsu's and Shi's works, a large sheet of Fermi surface predicted by theory (a large hole-like cubic sheet centered at $k = (\pi,\pi,\pi)$) was experimentally absent. Shi *et al.* have suggested that strong nesting between the flat faces of this sheet could have resulted in a charge density wave instability that results in this Fermi surface being entirely gapped. While a distinct possibility, such nesting driven density wave instabilities are highly uncommon in three-dimensional compounds due to phase space considerations, as pointed out by Chikamatsu *et al.*

At present, the origin of the apparent discrepancy between theory and experiment for these near E_F states in LSMO is still unclear. Because it is the states near E_F that are responsible for the thermodynamic properties, determining the electronic structure of these states and the suitability of theory to describe these materials remains a critical issue. In addition, LSMO is a polar compound, and charge redistribution at the surface which could alter the doping level on the surface relative to the bulk is another potential source of the discrepancy between experiment and theoretical predictions. At present, the possibilities regarding this apparent disagreement are as follows: whether this is due to (1) a failure of the density functional theory caused by the presence of strong electron correlations and magnetism, (2) the presence of a competing density wave instability which gaps out parts of the Fermi surface, (3) an altered surface electronic structure caused by the polar redistribution of charge and the surface termination, (4) strong electron–phonon interactions which dramatically suppress the near-E_F spectral weight resulting in highly massive polaronic charge carriers, or some combination of these various factors.

These first ARPES experiments have clearly demonstrated the importance and feasibility of thin film studies of LSMO and other closely related manganite perovskites. These measurements clearly demonstrate that the high energy structure of the valence band agrees well with predictions based on density functional theory, but clear discrepancies exist between the theory and experiment regarding the Fermi surface and near E_F band structure which are still unresolved. Therefore important questions remain which must be resolved before a clear understanding of the electronic structure of LSMO can be achieved through future ARPES experiments on LSMO thin films.

3.4 Future trends

The examples discussed in this chapter demonstrate the clear feasibility of performing *in situ* UPS on thin films. At present, we are only beginning to scratch the surface of what high-resolution angle-resolved UPS can achieve in revealing the electronic structure and many-body interactions in thin film samples. As described in this chapter, one major application of UPS and ARPES on *in situ* film samples is studying materials which do not have a natural cleavage plane, but which can be grown as atomically flat samples.

However, the next frontier in thin film materials is the design and discovery of new electronic materials at atomically abrupt interfaces between different correlated materials. While a great deal of work is currently underway in synthesizing interfacial electronic matter at oxide interfaces, these electronic states are extremely difficult to access using bulk probes. Hence, little is understood about the true electronic structure of these interfacial electronic states. ARPES and UPS are poised to play a key role in the study of *in situ* correlated interfaces, where the surface sensitivity of photoemission may be a double-edged sword. On the one hand, photoemission can be highly sensitive to the electronic states in a single monolayer, making it highly suitable for studying interfacial states. On the other hand, photoemission will have difficulty accessing interfaces which are deeply buried (more than 1–2 nm) below the surface. Along these lines, the development of more bulk-sensitive photoemission techniques, for instance by employing very low ($hv < 10\,\text{eV}$) or high ($hv > 1\,\text{keV}$) energy photon sources which increase the electron mean free path, may be critical for studying deeply buried interfaces. Bulk-sensitive photoemission will also be crucial for disentangling the surface and bulk electronic structures of materials which have a highly three-dimensional structure, like the examples provided in this chapter.

3.5 References

Berglund C N and Spicer W E (1964), 'Photoemission studies of copper and silver', *Phys. Rev. A*, **136**, 1030.

Borzi R A, Grigera S A, Farrell J, Perry R S, Lister S J S, Lee S L, Tennant D A, Maeno Y, and Mackenzie A P (2007), 'Formation of a nematic fluid at high fields in $Sr_3Ru_2O_7$', *Science*, **315**, 214.

Chikamatsu A, Wadati H, Kumigashira H, Oshima M, Fujimori A, Hamada N, Ohnishi T, Lippmaa M, Ono K, Kawasaki M, and Koinuma H (2006), 'Band structure and Fermi surface of $La_{0.6}Sr_{0.4}MnO_3$ thin films studied by *in situ* angle-resolved photoemission spectroscopy', *Phys. Rev. B*, **73**, 195105.

Damascelli A, Hussain Z, and Shen Z X (2003), 'Angle-resolved photoemission studies of cuprate superconductors', *Rev. Mod. Phys.*, **75**, 473.

Fujioka K, Okamoto J, Mizokawa T, Fujimori A, Hasi I, Abbate M, Lin H J, Chen C T, Takeda Y, and Takano M (1997), 'Electronic structure of $SrRuO_3$', *Phys. Rev B*, **56**, 6380.

Hufner S (1995), '*Photoelectron Spectroscopy: Principles and Applications*', New York, Springer-Verlag.

Kane E O (1964), 'Implications of crystal momentum conservation in photoelectric emission for band structure measurements', *Phys. Rev. Lett.*, **12**, 97–98.

Kiss T, Shimojima T, Ishizaka K, Chainani A, Togashi T, Kanai T, Wang X Y, Chen C T, Watanabe S, and Shin S (2008), 'A versatile system for ultrahigh resolution, low temperature, and polarization dependent laser-angle-resolved photoemission spectroscopy', *Rev. Sci. Inst.*, **79**, 023106.

Mackenzie A P and Maeno Y (2003), 'The superconductivity of Sr_2RuO_4 and the physics of spin-triplet pairing', *Rev. Mod. Phys.*, **75**, 657.

Mazin I I and Singh D J (1997), 'Electronic structure and magnetism in Ru-based perovskites', *Phys. Rev. B*, **56**, 2556.

Okamoto J, Mizokawa T, Fujimori A, Hase I, Nohara M, Takagi H, Takeda Y, and Takano M (1999), 'Correlation effects in the electronic structure of $SrRuO_3$', *Phys. Rev. B*, **60**, 2281.

Shi M, Falub M C, Willmott P R, Krempasky J, Herger R, Hricovini K, and Patthey L (2004), 'k-dependent electronic structure of the colossal magnetoresistive perovskite $La_{0.66}Sr_{0.34}MnO_3$', *Phys. Rev. B*, **70** 140407.

Siemons W, Koster G, Vailionis A, Yamamoto H, Blank D H A, and Beasley M R (2007), 'Dependence of the electronic structure of $SrRuO_3$ and its degree of correlation on cation off-stoichiometry', *Phys. Rev. B*, **76**, 075126.

Toyota D, Ohkubo I, Kumigashira H, Oshima M, Ohnishi T, Lippmaa M, Takizawa M, Fujimori A, Ono K, Kawasaki M, and Koinuma H (2005), 'Thickness-dependent electronic structure of ultrathin $SrRuO_3$ films studied by *in situ* photoemission spectroscopy', *Appl. Phys. Lett.*, **87**, 162508.

Traum M M, Smith N V, and DiSalvo F J (1974), 'Angular dependence of photoemission and atomic orbitals in the layer compound $1T-TaSe_2$', *Phys. Rev. Lett.*, **32**, 1241.

Xia J, Siemons W, Koster G, Beasley M R, and Kapitulnik A (2009), 'Critical thickness for itinerant ferromagnetism in ultrathin films of $SrRuO_3$', *Phys. Rev. B*, **79**, 140407(R).

4
X-ray photoelectron spectroscopy (XPS) for *in situ* characterization of thin film growth

H. BLUHM, Lawrence Berkeley National Laboratory, USA

Abstract: X-ray photoelectron spectroscopy (XPS) is an excellent tool for the investigation of the growth and reaction of thin films. Owing to the short mean free path of electrons in condensed matter, XPS is particularly well suited for the measurement of films with thicknesses of up to a few nanometers. XPS allows for the quantitative determining of the elemental composition, chemical specificity (i.e., oxidation state) and film thickness. In this chapter the basics of XPS are described, including different approaches to monitoring *in situ* film growth and reactions.

Key words: X-ray photoelectron spectroscopy (XPS), *in situ* techniques, thin films.

4.1   Introduction

X-ray photoelectron spectroscopy (XPS) is one of the most commonly used surface science techniques.[1–4] It is based on the photoelectric effect,[5] where an X-ray photon is absorbed by a core or valence electron. If the incident photon energy is larger than the binding energy of the electron, the electron will be emitted. The basic set-up of an XPS experiment is shown in Fig. 4.1. Monochromatic X-rays from a laboratory source (X-ray tube) or from a synchrotron irradiate the sample. The emitted electrons are collected by an electrostatic lens and energy-analyzed by a spectrometer, most commonly a hemispherical electron energy analyzer. From the known photon energy $h\nu$ and the measured kinetic energy KE, the binding energy BE of the electrons can be determined according to BE = $h\nu$–KE–Φ, with Φ as the spectrometer work function. Typical incident photon energies in an XPS experiment range from several 10s to well over 1000 eV, with kinetic energies of the same order.

While the penetration depth of the X-rays into the sample is of the order of hundreds of nanometers or more,[6] elastic and inelastic interactions of photoelectrons with atoms in the sample limit the probed depth to a maximum of a few nanometers over the kinetic energy range used in conventional XPS studies. Figure 4.2 shows the inelastic mean free path (IMFP) of electrons as a function of their kinetic energy in Cu; the minimum path length is at about 100 eV kinetic energy (see Fig. 4.2).[7] This surface sensitivity makes

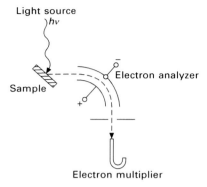

4.1 Basic set-up of an XPS experiment. The sample is irradiated by monochromatic X-rays. The kinetic energy of the emitted photoelectrons is analyzed by an electron analyzer. Reprinted with permission from Ertl and Küppers.[3] Copyright (1985), Wiley-VCH Verlag GmbH & Co. KGaA.

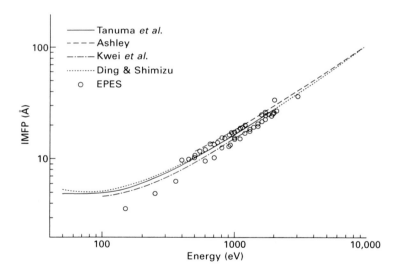

4.2 Inelastic mean free path as a function of kinetic energy of electrons in copper. The graph shows calculated data from: Tanuma et al., *Surf. Interface Anal.*, **17**, 911 (1991); Ashley, *J. Electron Spectrosc. Rel. Phenom.*, **50**, 323 (1990); Kwei et al., *Surf. Sci.*, **293**, 202 (1993); as well as Ding and Shimizu, *Scanning*, **18**, 92 (1996). Experimentally determined IMPF values from elastic peak electron spectroscopy measurements and Monte Carlo simulations are also shown (Powell and Jablonski, *J. Phys. Chem. Ref. Data*, **28**, 19 (1999)). Reprinted with permission from Powell and Jablonski.[8] Copyright (2009), Elsevier BV.

XPS particularly well suited for the study of ultrathin films with only a few monolayer or less in thickness. Figure 4.2 also shows that the IMFP increases with increasing kinetic energy, e.g., in the case of Cu up to about 10 nm at a kinetic energy of 10 keV. With the use of high-brilliance synchrotrons that provide X-rays in this energy range, the study of thick multilayer system and buried interfaces has become feasible, as will be discussed later in this chapter. It should be noted that the inelastic mean free path is not a precise description of the actual probing depth in an XPS experiment, since it neglects elastic scattering effects. A more accurate quantity for the measurement depth is the effective attenuation length (EAL), which includes elastic scattering effects.[8]

While the short mean free path of electrons in matter makes XPS an exquisitely surface-sensitive method, it also requires in general high vacuum conditions during the measurements, since the emitted photoelectrons are also scattered by gas phase molecules. For example, the mean free path of 100 eV electrons (i.e., those with the highest surface sensitivity) is about 1 mm in 1 torr of water vapor. Since the electrons travel many centimeters on their way to the electron detector, attenuation of the signal under elevated pressure conditions poses a limit on the background pressure in conventional XPS set-ups. Over the years, different schemes for *in situ* XPS measurements have been developed that now allow measurements at pressures in the torr range, as will be shown later in the chapter.

The advantages of XPS for the investigation of thin films are its surface sensitivity and chemical specificity. Since each element exhibits a characteristic set of core level peaks in XPS, the chemical composition of the surface and near-surface region of a sample can be determined quantitatively. Moreover, the binding energy of core and valence electrons shows subtle changes (of the order of tenths to several eV) depending on the chemical bonding (e.g., the oxidation state). This allows the discrimination, for instance, of Si atoms in an oxide layer ($\sim$Si^{4+}) from Si atoms at the interface ($\sim$Si^{1+}, Si^{2+}, Si^{3+}) and elemental Si in the substrate (Si^0) of a Si wafer covered by a native oxide layer (see Fig. 4.3).[9] When interpreting these so-called 'chemical shifts', one has to keep in mind that the binding energy depends both on the state of the atom before the photoemission event ('initial state effects') as well as contributions due to the reorganization of the electrons in the atom during the relaxation process, when the core hole is filled ('final state effects'). It is, however, in many cases possible to assign chemical shifts to initial state effects. For instance, the observed chemical shifts for the different Si species in Fig. 4.3 can be rationalized by different degrees of charge transfer from Si to O (depending on the Si oxidation state), which leads to variations in the screening of the Si 2p electrons by Si 3s and 3p valence electrons. Thus, the 2p electrons in the case of Si^{4+} experience a stronger Coulomb attraction to the nucleus than in Si^{3+}, resulting in a higher binding energy in the case of

78 *In situ* characterization of thin film growth

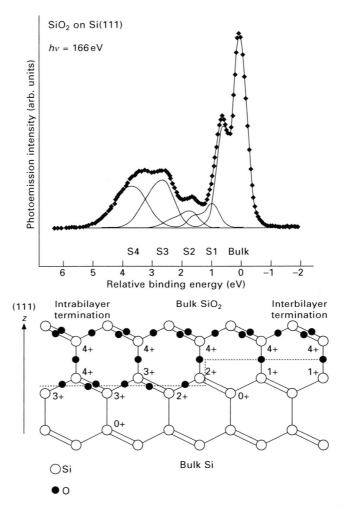

4.3 Upper panel: Si 2p XPS spectrum for $SiO_2/Si(111)$. The diamonds are data points, and the curves show a fit and the decomposition into components. Lower panel: model for the $SiO_2/Si(111)$ interface. Two possible terminations are shown. The labels indicate the oxidation states for Si atoms at and near the interface. Reprinted with permission from Sieger *et al.*[9] Copyright (1996), by the American Physical Society.

Si^{4+}. This sensitivity to the chemical state of the atoms means that valuable information can be gleaned from XPS measurements about, e.g., bonding within thin films, at the surface, as well as at the interface to the substrate (see Fig. 4.3).

XPS can also be used to determine the thickness of deposited layers on a substrate, assuming ideal layer-by-layer growth and smooth interfaces. For a single layer with homogeneous chemical composition and thickness d_m, the detected intensity I_m, using the straight-line approximation, is given by[2,10]

$$I_m = S_m^j \int_0^d e^{-z/\lambda'_m} dz = S_m^j \lambda'_m (1 - e^{-d_m/\lambda'_m}) \qquad [4.1]$$

with $\lambda'_m = \lambda_m \cos(\alpha)$ as the effective attenuation length at angle α between the surface normal and detector position (see Fig. 4.4), λ_m the attenuation length of electrons for layer m at a given kinetic energy, and z the depth measured from the layer surface. The factor S contains a number of experimental parameters and is given for layer m and inner shell orbital j by:

$$S_m^j = \Phi_m(hv) \times \sigma_m^j(hv) \times \beta_m^j(hv) \times D_m(\text{KE}) \times N_m \qquad [4.2]$$

with Φ as the X-ray flux at a given X-ray energy (hv), σ the photoemission cross-section[11] at a given hv, β the orbital specific asymmetry parameter, D the spectrometer efficiency for a given kinetic energy, and N the number of atoms per unit volume. For an infinitely thick layer with $d_m \to \infty$ (i.e., a bulk substrate), the intensity is $I_\text{subs} = S_\text{subs}^j \lambda'_\text{subs}$.

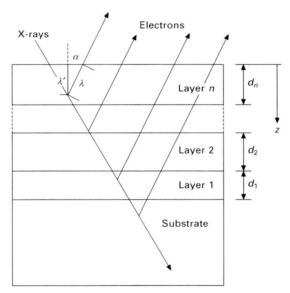

4.4 Model for a multilayer system consisting of n homogeneous layers above a bulk substrate, where the order of stacking from bottom to top is given by layer 1, layer 2, ..., layer n. The detection angle of the electron analyzer with respect to the surface normal is α. The effective probing depth λ' is given by the electron attenuation length λ times $\cos(\alpha)$.

For a multilayer system consisting of n homogeneous layers above a bulk substrate (see Fig. 4.4), where the order of stacking from bottom to top is given by layer 1, layer 2, ..., layer n, the XPS intensity from the substrate is attenuated by each layer according to the Beer–Lambert law:[12]

$$I_{\text{subs}} = S_{\text{subs}}^{j} \lambda'_{\text{subs}} \prod_{i=1}^{n} e^{-d_i/\lambda'_i} \qquad [4.3]$$

Likewise, the intensity of layer m in a system of n layers is being attenuated by $(n - m)$ layers, and is thus given by:

$$I_m = S_m^{j} \lambda'_m (1 - e^{-d_m/\lambda'_m}) \prod_{i=m+1}^{n} e^{-t_i/\lambda'_i} \qquad [4.4]$$

There are two principal methods to obtain depth-dependent elemental and chemical information using XPS: (1) depth-profiling by etching (for instance with argon ions), where surface layers are gradually removed by sputtering, while XPS spectra are obtained at various stages during the removal of surface layers, and (2) non-destructive depth profiling through the variation of the escape-depth of the electrons, either by varying the detection angle relative to the sample surface or by measuring photoelectrons with different kinetic energies. The advantage of depth profiling by argon etching is that it is not limited by the escape depth of the electrons, so that information about the sample from depths in excess of several nm can be obtained. The disadvantage is that the sample surface is destroyed in the process, and that the chemical state of the surface may be altered due to the impact of Ar ions.[13]

Non-destructive depth profiling in an XPS experiment that uses an X-ray source with a fixed energy (i.e., a lab-based experiment) relies on the variation of the take-off angle α of the photoelectrons (see Fig. 4.4), which determines (at a fixed kinetic energy) the information depth in the experiment.[2,14] An example for this technique is shown in Fig. 4.5 for the case of a 4 nm thick SiO_xN_y film on a Si substrate.[15] The relative intensity of the Si^{4+} peak is clearly enhanced at shallow detection angle. The schematic representation of the electron trajectories in Fig. 4.5 also illustrates that the depth resolution in angle-resolved XPS depends on the angular acceptance of the electrostatic analyzer, which should be ideally as narrow as possible. Since the effective probing depth λ' varies with the take-off angle α according to $\lambda \cos(\alpha)$ (with λ as attenuation length), the most sensitive determination of the film thickness is at the most grazing detection angles. Measurements at these angles, however, are hampered by surface roughness effects, which need to be taken into account.[14]

The use of variable incident photon energies (and thus kinetic energies) for depth profiling is performed using a fixed sample–detector geometry and is thus not influenced by surface geometry effects. This method requires a

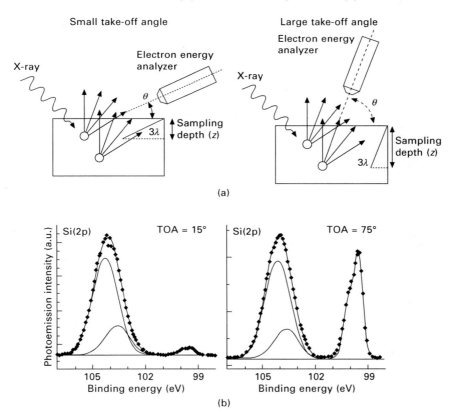

4.5 (a) Schematic depiction of an angle-resolved XPS experiment for the measurement of the chemical states and composition of SiO$_x$N$_y$ films on a Si substrate. (b) Si 2p XPS spectra of a 4 nm SiO$_x$N$_y$ film, taken at 15° and 75° take-off angles. The relative intensity of Si^0 to Si^{4+} is enhanced with increasing take-off angle (TOA). Reprinted with permission from Chang et al.[15] Copyright (2000), American Institute of Physics.

photon source with variable energy, such as a synchrotron. The technique makes use of the dependence of the mean free path of electrons in matter on their kinetic energy (see Fig. 4.2). An example is shown in Fig. 4.6, where the growth of praseodymium-SiO$_2$ layers on Si was monitored using XPS.[16] Figure 4.6(a) shows Si 2p spectra after each preparation step: first after the growth of a 2.6 nm thick SiO$_2$/Si film, then after the deposition of 1 nm of metallic Pr on top of SiO$_2$, and finally after annealing the sample to 600 °C, which leads to incorporation of Pr into SiO$_2$ in a solid state reaction. Under each condition Si 2p spectra were taken with kinetic energies of 440, 660 and 920 eV, with the most surface-sensitive spectrum at 440 eV and the most

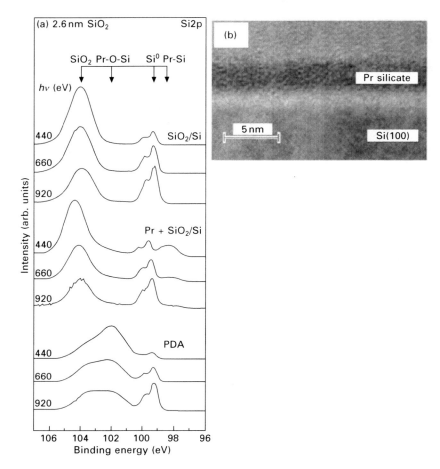

4.6 (a) Si 2p spectra taken at different photoelectron kinetic energies during the preparation of Pr-SiO$_2$ layers on a Si substrate. (b) Transmission electron microscopy cross-sectional images of the film shown in (a). Reprinted with permission from Lupina *et al.*[16] Copyright (2005), American Institute of Physics.

bulk sensitive at 920 eV. The spectra after annealing show a strong peak at a binding energy of ~102 eV in addition to the Si^0 and Si^{4+} peaks due to the substrate and the SiO$_2$ film, respectively. The peak at 102 eV is attributed to the formation of Pr silicate, which is located predominantly at the sample surface, as the depth profiling spectra after post-deposition annealing (or PDA, in Fig. 4.6a) prove. This scenario is confirmed by cross-sectional transmission electron microscopy images that were taken after the reaction (Fig. 4.6b).

To summarize so far, XPS is a quantitative method for the determination of thin film elemental composition and chemistry and can also be used under favorable circumstances to determine the film thickness, as well as to distinguish surface from bulk species. The following section discusses how XPS can be used to *in situ* monitor thin film growth, the reaction of thin film surfaces with adsorbates and gases, as well as the chemistry of solid/solid interfaces in multilayer systems. Some of these investigations required the development of new XPS instrumentation, which will be discussed alongside the experimental results.

4.2 *In situ* monitoring of thin film growth

Studying the elemental and chemical composition as well as the thickness of thin films *in situ* during film growth using XPS requires an expansion of the technique to pressures in the mtorr range. The first dedicated XPS instrument for monitoring thin film growth was described by Kelly *et al.* in 2001.[17] A schematic of the instrument is shown in Fig. 4.7. This system is capable of operating at pressures in the 10^{-3} torr range, about three orders of magnitude higher than the pressure limit in conventional XPS systems. The X-ray source (Mg Kα anode) is separated from the gas atmosphere in the measurement chamber by a 2 µm thick aluminum foil, which is about 70% transparent for Mg Kα radiation (1254 eV).[6] The electrostatic lens system of a commercial hemispherical analyzer was modified by introducing a 1 mm diameter differentially pumped aperture in the lens column. Two spherical stainless steel meshes are mounted in front of the standard input lens to collect and focus electrons from the sample onto the differentially pumped aperture (see Fig. 4.8). After passing through the meshes the electrons are accelerated to energies between 1000 to 1300 eV, which decreases the lens magnification to about 1 and also decreases the scattering of electrons by gas molecules due to the high electron kinetic energy. At the given lens magnification the electron analyzer measures a sample area of about 1 mm diameter.

The pressure differential between the measurement/thin film growth chamber and the hemispherical analyzer was 3–4 orders of magnitude, providing the necessary pressure difference for operation at mtorr pressures in the experimental chamber. Other considerations in the design of the spectrometer included high-speed acquisition of spectra to follow film growth with high-time resolution, a large working distance (~3 to 5 cm) between the electrostatic analyzer input lens and the sample to avoid interference by the electrostatic lens system with the reactant flow to the sample, as well as the suppression of stray electrons and ions that may be generated in plasmas near the growing surface.[17] The latter is achieved by biasing the meshes in the input lens positively at 20 V (thus rejecting slow ions) and operating the lens elements in a manner that reduces the transmission of secondary electrons. High data acquisition speed was achieved by a large collection

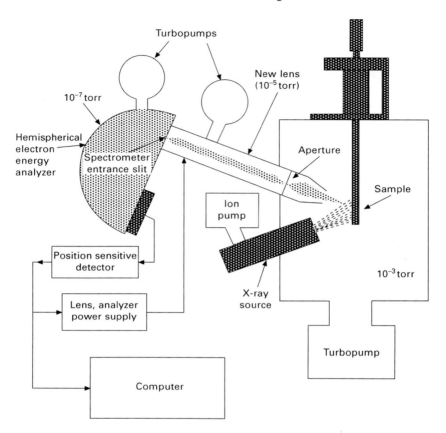

4.7 Set-up of an XPS spectrometer for the *in situ* monitoring of thin film layer deposition. A conventional Mg Kα X-ray source is separated from the elevated pressure in the experimental chamber by a 2μm thick Al foil. The differentially pumped electron lens focuses emitted photoelectrons through a small aperture with a two orders-of-magnitude pressure differential across it. The electrons are then imaged onto the spectrometer entrance slit into a still lower pressure region. Reprinted with permission from Kelly *et al.*[17] Copyright (2001), American Vacuum Society.

angle of the electrostatic lens (30° total) and the use of an imaging electron detector that measures a range of electron energies simultaneously.

A case study for the application of this instrument is shown in Fig. 4.9. The growth of a thin film of tungsten oxide on a Si substrate was monitored by measuring the W 4f, O 1s and Si 2p core level peaks (see upper panel in Fig. 4.9). The acquisition time for each spectrum is 30 s. The initial substrate surface shows the characteristic Si 2p peaks for a native oxide layer on a Si substrate, where the peak in the O 1s spectrum is purely due to the native

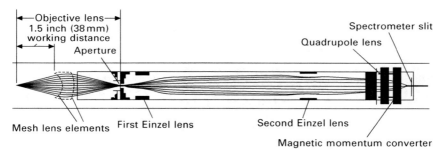

4.8 Detailed schematic view of the electrostatic lens system used in the spectrometer shown in Fig. 4.7. Reprinted with permission from Kelly et al.[17] Copyright (2001), American Vacuum Society.

silicon oxide layer. Tungsten is evaporated onto the substrate from a hot filament in a background of 1 mtorr of oxygen. The evolution of the W 4f, O 1s and Si 2p peaks during deposition shows a decrease in both the Si and SiO_x peaks due to attenuation of Si photoelectrons by the deposited layer. As expected, the W 4f signal increases with increasing deposition time. The same holds for the O 1s intensity, which indicates that oxygen is incorporated into the growing film, since the initial oxygen intensity was only due to oxygen in SiO_2, and the Si 2p peak of SiO_2 decreases with time. The integrated intensities for Si 2p, O 1s and W 4f as a function of deposition time are plotted in the bottom panel of Fig. 4.9.

The Si 2p and W 4f spectra at deposition times of ~11 min ('A') and ~61 min ('B') are displayed in Fig. 4.10, upper panels.[17] The Si 2p spectra show a decrease in intensity in both the SiO_2 and the elemental Si 2p peaks, as expected for the growth of a layer at the surface. The W 4f spectra initially show the characteristic binding energy for a WO_3 species ('A'), while at later deposition times some elemental W is also observed. By relating the Si 2p intensity of the pristine substrate to the Si 2p intensity at various stages during deposition, the attenuation of the Si 2p signal can be used to determine the thickness of the WO_x film (see Eq. [4.3]). With the detection angle at 25° from the sample normal, and assuming an attenuation length of 1.8 nm for the Si 2p electrons (kinetic energy ~ 1150 eV), the escape depth of the electrons in the experiment is 1.8 nm × cos(25°) ~ 1.6 nm. Using these values, a model for the thickness of the SiO_2 and WO_x layers at different stages during the deposition can be postulated (bottom panel of Fig. 4.10).

Apart from the experiments shown in Figs 4.9 and 4.10, this instrument has also been used to monitor the deposition of Al_2O_3 on InGaAs,[18,19] and the oxidation of Ge(100).[20] While the performance of the current set-up used in the case studies described above provided hitherto unavailable

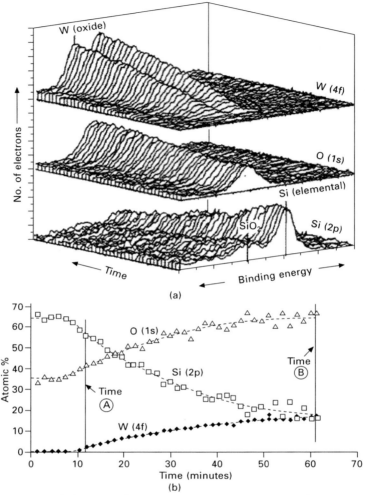

4.9 (a) W 4f, O 1s and Si 2p spectra obtained *in situ* during the growth of a WO$_x$ film on a SiO$_2$/Si substrate using the spectrometer shown in Fig. 4.7. (b) Integrated XPS peak intensities plotted as a function of time. Reprinted with permission from Kelly *et al.*[17] Copyright (2001), American Vacuum Society.

opportunities for *in situ* film growth monitoring, Kelly *et al.* estimated that the performance of the instrument can be increased by more than an order of magnitude through two measures: (a) increasing the acceptance angle of the spectrometer to 40°, and (b) by use of a state-of-the-art focused X-ray source instead of the unfocused one which is used in the current set-up. A focused X-ray source would increase the X-ray flux in the 1 mm^2 field of view of the detector by more than a factor of 10. This would then allow the

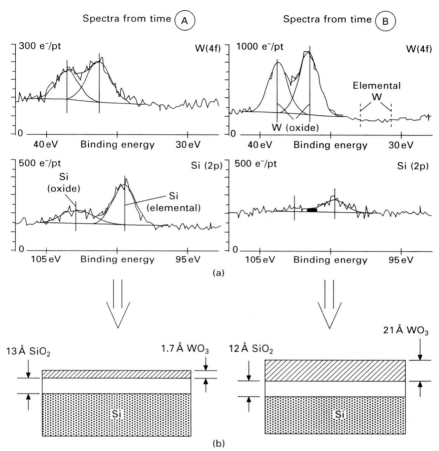

4.10 (a) Detailed Si 2p and W 4f spectra taken after 11 (A) and 61 minutes (B) during the growth of a WO_x film on SiO_2/Si substrate (e⁻/pt = number of electrons per unit (point) on *y*-axis). (b) A model for the evolution of the WO_x film with time can be postulated from the integrated peak intensities as well as the probing depth in the experiment. Reprinted with permission from Kelly *et al.*[17] Copyright (2001), American Vacuum Society.

acquisition of the order of one spectrum per second, an improvement of a factor >10 over the current speed. If in addition the lens design were modified so that the acceptance area of the lens is increased to 4 mm², another factor of 10 in acquisition speed could be gained, making it possible to collect 10 spectra per second. Even though these improvements are challenging, they are merely technical in nature; monitoring film growth with a time resolution of 0.1 s should therefore be possible in the future.

4.3 Measuring the reaction of thin films with gases using ambient pressure X-ray photoelectron spectroscopy (XPS)

In the following we will focus on the reaction of gases with thin films. Ultrathin oxide, metal, organic as well as semiconductor films play an ever-increasing role in technological applications, including industrial catalysis and data storage devices. The interaction of these systems with reactants as well as environmental gases (such as water vapor) has great influence on their performance and longevity. In particular in the field of heterogeneous catalysis the correlation between the yield and conversion in a catalytic reaction (measured by the composition of the gas phase) with the chemical nature of the active catalyst surface is of great importance for a better understanding of the basic, atomic scale processes in a catalytic reaction, and may lead to a rational design of more efficient catalytic materials.

On the other hand, thin film systems can also be used to study surface reactions, in particular phase transitions and volatilization processes[21] that are difficult to quantify in a bulk system. This is illustrated in Fig. 4.11. The upper panel shows the scenario for a bulk sample that interacts with the gas environment and forms a reacted layer at its surface. Photoelectrons with a certain escape depth (symbolized by the length of the arrow) are used to monitor the reaction. If the reaction also involves volatilization, this process

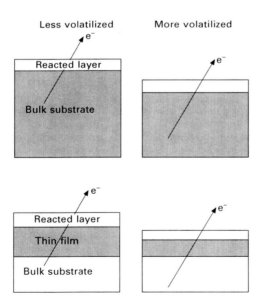

4.11 Volatilization and phase transition reactions are easier to quantify in a thin film system than for a bulk sample. For details see text.

is not detectable in an experiment on a bulk sample: the sample on the left will give an identical signal in an XPS experiment as the sample on the right in the upper panel. However, if one uses a thin film sample instead (lower panel, Fig. 4.11) where the film thickness is of the order of the escape depth of the electrons, the attenuation of the substrate electrons can be used to gauge the degree of volatilization of the film, or of its partial conversion into another phase. In addition, since the sample under investigation has a finite thickness of the order of the escape depth of the electrons, the signal of the sample itself can be used to measure the volatilization or conversion of the material into another phase. In essence, thin film samples are, under certain circumstances, superior to bulk samples to monitor gas/surface interactions. It is, however, necessary to point out two caveats. It has been shown that thin film systems can show markedly different properties from their bulk counterparts; this is in part their appeal for the tuning of reaction properties in catalysis.[22] In addition, all the above considerations hold only true in the absence of morphological changes (i.e. deviations from a strictly two-dimensional model) to the film and substrate–film interface; such changes would make the quantitative analysis of thickness changes challenging.

The investigation of the reaction of surfaces with gas phase species needs to bridge the so-called 'pressure gap' in surfaces science. In the case of XPS this is hampered by the strong interaction of electrons with gas phase molecules, as pointed out in the section above. The differentially pumped electrostatic lens designed by Kelly *et al.* afforded measurements at pressures in the mtorr range.[17] Many reactions, in particular in environmental science, require higher pressures: in order to measure, e.g., the surface of neat liquid water the water vapor pressure in the experimental chamber has to be at least 4.6 torr, which is the equilibrium water vapor pressure at the triple point.

To achieve higher pressures in an XPS experiment, the path length of the electrons through the high-pressure region has to be kept as short as possible. In addition, several differential apertures are necessary to keep the electron analyzer in a high-vacuum environment. This basic concept (see Fig. 4.12a and b) was developed more than 30 years ago in the original designs by Hans & Kai Siegbahn and collaborators, which allowed experiments of up to 1 torr.[23,24] Several other groups built instruments based on this concept.[25–27] To overcome the trade-off between an increase in detection efficiency through larger apertures on one hand, and better differential pumping through smaller apertures on the other hand, the latest generation of these instruments uses electrostatic lenses that are placed between the apertures, raising the pressure limit to more than 5 torr.[28–31] The increase in the pressure limit is also partly due to the use of synchrotron radiation, which offers higher photon flux and tighter focused X-ray beams. Since the instruments operate at realistic environmental humidities, the technique is often called ambient pressure

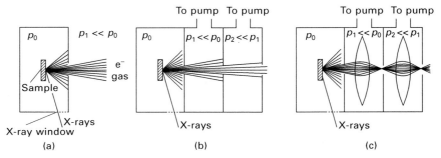

4.12 Principle of APXPS instruments. (a) The sample is mounted in a high-pressure chamber, close to a differentially pumped aperture through which electrons and gas escape. The X-ray source is separated from the high-pressure cell by an X-ray-transparent window. (b) To maintain a vacuum better than 10^{-6} torr for typical entrance aperture diameters of ~1 mm and pressures in the torr range in the sample cell, several differential pumping stages are needed. There is a trade-off between the pumping efficiency, which increases with aperture spacing and decreasing aperture sizes, and the collection efficiency of electrons, which is determined by the solid angle that is subtended by the apertures. (c) In a differentially pumped electrostatic lens system the electrons are focused onto the apertures between the differential pumping stages, thus increasing the transmission of electrons through the differential pump stages and allowing for reduced aperture sizes for improved differential pumping. Reprinted from Bluhm.[34] Copyright (2010), with permission from Elsevier.

XPS (APXPS). Review articles on the basics and applications of APXPS can be found in Refs. 29, 32, 33 and 34.

The application of APXPS to the study of the reaction of thin films is now illustrated using the example of the interaction of a 4 monolayer (ML) thick MgO(100) film grown on a Ag(100) substrate with water vapor. Due to its simple rock salt structure, MgO(100) is an ideal model metal oxide surface for studying the metal oxide–water interface, both theoretically and experimentally.[35] Of special interest is the nature of the interaction of water vapor with MgO, in particular molecular vs dissociative adsorption. In this case study, MgO(100) was grown on Ag(100) by vapor deposition following the method reported by Wollschläger *et al.*[36] Magnesium was deposited at a rate of ~0.1 nm min^{-1} in the presence of 10^{-6} torr O_2 while maintaining the Ag(100) substrate at 300 °C. The interaction of water with the MgO(100) film surface was then studied by monitoring the O 1s and Ag 3d spectra in isobar experiments, where the water vapor pressure was kept constant while the sample temperature was decreased from well over 300 °C to below room temperature.[37] Figure 4.13 shows O 1s, Ag 3d, as well as C 1s spectra taken in a 0.15 torr isobar experiment. The O 1s spectrum exhibits four peaks

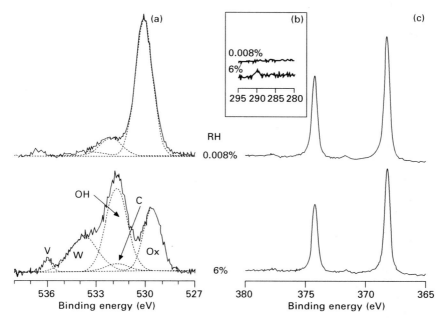

4.13 Ambient pressure XPS O 1s (a), C 1s (b) and Ag 3d (c) spectra taken at a relative humidity of 0.008% and 6% on a 4 ML thick MgO(100)/Ag(100) film. All spectra are taken at an electron kinetic energy of 220 eV. Peak designations in the O 1s spectrum are the metal oxide (Ox), hydroxyl (OH), surface molecular water (W), water vapor (V), and OH contribution due to oxidized carbon species (C). Reprinted from Newberg *et al.*[37] Copyright (2011), with permission from Elsevier.

assigned to (from low to high binding energy) oxide, hydroxide, adsorbed molecular water, and water vapor (the gas phase in front of the sample is also partly detected in the experiment). A comparison of the O 1s spectrum at a relative humidity of 0.008% with the one at 6% shows major changes in the MgO film: there is a strong increase of hydroxide uptake as well as water adsorption at higher humidity. The Ag 3d spectra do not show any change (apart from attenuation), which indicates that the MgO film is covering the whole of the Ag(100) substrate, thus preventing a reaction between water vapor and Ag substrate. In addition, C 1s spectra do show a slight increase in carbonaceous species, which is expected under ambient measurement conditions.[34]

The O 1s oxide, hydroxide and adsorbed water peak intensity can be converted into equivalent monolayer coverages using a procedure (see Newberg *et al.* 37) that analyzes both the decrease in Ag 3d signal due to the adsorption of OH and H_2O, as well as the relative O 1s signals from the film. The results are plotted in Fig. 4.14. There is a sharp onset of surface

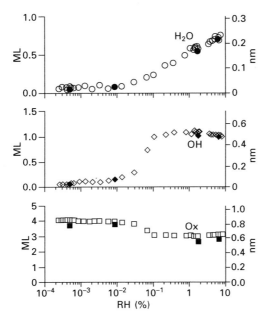

4.14 Integrated O 1s peak intensities from ambient pressure XPS 0.15 torr isobar measurements on 4 ML MgO(100)/Ag(100). Peak assignations are as in Fig. 4.13. The film thicknesses are plotted in units of monolayers (left axis) and nanometers (right axis). 1 ML of Ox, OH, and water are defined as 0.21, 0.48 and 0.31 nm, respectively. Reprinted from Newberg et al.[37] Copyright (2011), with permission from Elsevier.

hydroxylation at a relative humidity of ~0.01%, accompanied by a reduction in oxide coverage. Water adsorption increases notably at this relative humidity, too. A detailed analysis of the data leads to a model for the reaction of water vapor with MgO(100) under ambient relative humidity, depicted in Fig. 4.15. At a critical relative humidity (~0.01%), water molecules dissociate on the MgO(100) surface, leading to the hydroxylation of the oxygen atoms in the topmost MgO layer, and to the addition of a monolayer of OH groups to the surface. The validity of this reaction mechanism is borne out in the reduction of the total MgO(100) film thickness and the increase in the total O 1s surface signal. This underscores the utility of the application of thin films for the study of surface reactions, as depicted in Fig. 4.11.

4.4 *In situ* measurements of buried interfaces using high kinetic energy XPS (HAXPES)

In the final part of this chapter we would like to illuminate another promising XPS technique for the *in situ* study of thin film properties, namely high

X-ray photoelectron spectroscopy (XPS)

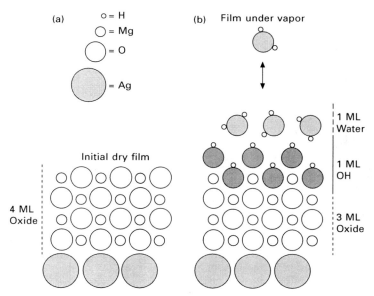

4.15 Illustration of the reaction of a MgO(100)/Ag(100) film with water vapor. (a) Without and (b) with water vapor. In the presence of water vapor at a relative humidity larger than 0.01%, water molecules at the surface dissociate to form two hydroxyl ion moieties: those sitting above the initial MgO interface near Mg^{2+} surface sites, and those that convert the top layer of MgO oxygen ions to OH^- ions. Assuming the entire MgO interface is passivated with OH, the net effect reduces the MgO film from 4 to 3 ML. Reprinted from Newberg et al.[37] Copyright (2011), with permission from Elsevier.

kinetic energy XPS. This technique has been pioneered by Lindau et al. who demonstrated in 1974 the use of high-energy photons to measure the intrinsic line width of Au 4f core levels.[38] The attraction of high kinetic energy XPS lies in the ability to measure buried interfaces at much larger depths than in conventional soft X-ray XPS (close to 10 nm at 10 keV in the case of Cu, as opposed to close to 1 nm at 1 keV – see Fig. 4.2). However, the detrimental aspects of this technique are the low photoemission cross-sections at high kinetic energies (several orders of magnitude lower than at low kinetic energies) as well as the need for high voltages on the lens elements for sufficient deceleration of electrons for high energy resolution measurements. With the recent availability of high-brightness synchrotron sources the low cross-sections can be compensated for. Figure 4.16 shows recent results for buried SiO_2 layers on a Si substrate, which are covered by a NiGe overlayer. Using incident photons with an energy of 7935 eV the authors recorded Si 1s spectra (binding energy ~1840 eV) with electron kinetic energies of over 6000 eV, which allowed probing at a depth of more

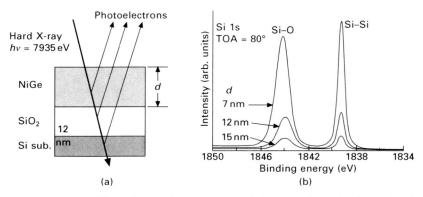

4.16 (a) Experimental parameters of the sample under investigation. (b) Si 1s XPS spectra of the sample shown in (a). The incident photon energy is 7935 eV. Spectral signatures from the SiO$_x$ film and the Si substrate can be observed even at a NiGe overlayer thickness of 15 nm. Reprinted from Kobayashi.[39] Copyright (2009), with permission from Elsevier.

than 15 nm.[39] In another case study the smoothness of the solid/solid interface of buried Ni/Cu interfaces was probed using high kinetic energy XPS.[40] Figure 4.17 (upper panel) shows the layout of the sample. Alternate layers of nickel and copper (5 ML each) were deposited on a MgO(100) substrate that was covered by intermediate layers of Fe, Pt, Cu and Ni. This structure was covered by a 1 nm thick Pt cap layer. High kinetic energy (~1200 eV) XPS spectra with a resolution of ~0.25 eV were taken of the Cu 2p core level (bottom panel of Fig. 4.17). Upon heating of the sample from room temperature to 300 °C a marked shift in the Cu $3p_{3/2}$ peak position to lower binding energy was observed, which was interpreted as a signature for the smoothing of the Cu/Ni interface.[41] These results demonstrate that XPS is not only a valuable tool for the study of the sample surface or near-surface region, but also for investigation of solid/solid interfaces up to a depth of several nanometers.

4.5 Conclusions

In summary, X-ray photoelectron spectroscopy is an excellent tool for the investigation of *in situ* film growth, reaction, as well as the study of buried interfaces. Recent developments of *in situ* XPS spectrometers as well as high-brilliance synchrotron facilities have expanded the application of XPS to higher pressures as well as larger sampling depths. The challenge for the future lies in the adaption of XPS instruments to realistic environments of thin film growth as well as gas/solid interactions.

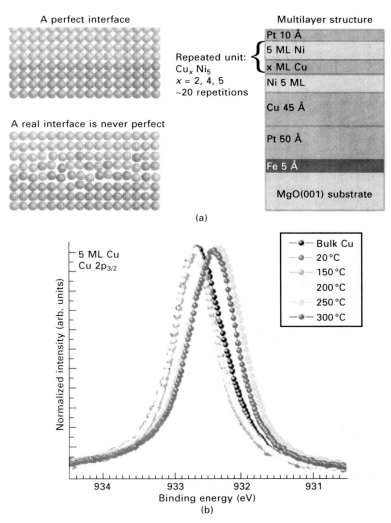

4.17 (a) Left: schematic view of an idealized (top) Ni/Cu interface, where the interfacial roughness is negligible, as compared to a realistic interface (bottom). The right panel shows the configuration of the sample under investigation. (b) Cu $2p_{3/2}$ spectra, taken at an incident photon energy of 2010 eV. The shift in the core level binding energy at temperatures above 250 °C indicates a change in the roughness of the Ni/Cu interface. Reprinted from Gorgoi *et al.*[40] Copyright (2009), with permission from Elsevier.

4.6 Acknowledgments

The contributions in particular of D. E. Starr, J. T. Newberg, E. R. Mysak, and K. R. Wilson are gratefully acknowledged. This research, the ALS, and the ALS-MES beamline 11.0.2 are supported by the Director, Office of Science, Office of Basic Energy Sciences, Division of Chemical Sciences, Geosciences of the US Department of Energy at the Lawrence Berkeley National Laboratory under Contract No. DE-AC02-05CH11231.

4.7 References

1. S. Hüfner, *Photoelectron Spectroscopy*, 3rd Ed., Springer Verlag, Berlin, 2003.
2. D. Briggs, M.P. Seah, *Practical Surface Analysis, Volume 1 – Auger and X-ray photoelectron spectroscopy*, 2nd Ed., John Wiley & Sons, Chichester, 1996.
3. G. Ertl, J. Küppers, *Low Energy Electrons and Surface Chemistry*, 2nd Ed., VCH Verlagsgesellschaft mbH, Weinheim, 1985.
4. J.W. Niemantsverdriet, *Spectroscopy in Catalysis*, 2nd Ed., Wiley VCH, Weinheim, 2000.
5. A. Einstein, Über einen die Erzeugung und Verwandlung des Lichtes betreffenden heuristischen Gesichtspunkt, *Ann. Physik* **17**, 132 (1905).
6. Center for X-ray Optics at Lawrence Berkeley National Laboratory (Berkeley, CA); http://henke.lbl.gov/optical_constants/.
7. M.P. Seah, W.A. Dench, Quantitative electron spectroscopy of surfaces: a standard data base for electron inelastic mean free paths in solids, *Surface Interface Anal.* **1**, 2 (1979).
8. C.J. Powell, A. Jablonski, Surface sensitivity of X-ray photoelectron spectroscopy, *Nucl. Instr. Meth. A* **610**, 54 (2009).
9. M.T. Sieger, D.A. Luh, T. Miller, T.-C. Chiang, Photoemission extended fine structure study of the $SiO_2/Si(111)$ interface, *Phys. Rev. Lett.* **77**, 2758 (1996).
10. C.S. Fadley, Angle-resolved X-ray photoelectron spectroscopy, *Prog. Surf. Sci.* **16**, 275 (1984).
11. J.J. Yeh, I. Lindau, Atomic subshell photoionization cross sections and asymmetry parameters: $1 \leq Z \leq 103$, *At. Data Nucl. Data Tables* **32**, 1 (1985).
12. R. W. Paynter, An ARXPS primer, *J. Electron. Spectrosc. Relat. Phenom.* **169**, 1 (2009).
13. D. Briggs, *Surface Analysis of Polymers by XPS and Static SIMS*, Cambridge University Press, Cambridge, 1998.
14. R.J. Baird, C.S. Fadley, X-ray photoelectron angular distributions with dispersion-compensating X-ray and electron optics, *J. Electron. Spectrosc. Relat. Phenom.* **11**, 39 (1977).
15. J.P. Chang, M.L. Green, V.M. Donnelly, R.L. Opila, J. Eng, Jr., J. Sapjeta, P.J. Silverman, B. Weir, H.C. Lu, T. Gustafsson, E. Garfunkel, Profiling nitrogen in ultrathin silicon oxynitrides with angle-resolved X-ray photoelectron spectroscopy, *J. Appl. Phys.* **87**, 4449 (2000).
16. G. Lupina, T. Schroeder, J. Dabrowski, Ch. Wenger, A. Mane, G. Lippert, H.-J. Müssig, P. Hoffmann, D. Schmeisser, Praesodymium silicate layers with atomically abrupt interface on Si(100), *Appl. Phys. Lett.* **87**, 092091 (2005).
17. M.A. Kelly, M.L. Shek, P. Pianetta, T.M. Gür, M.R. Beasley, In situ X-ray photoelectron spectroscopy for thin film synthesis monitoring, *J. Vac. Sci. Technol. A* **19**, 2127 (2001).
18. B. Shin, J.B. Clemens, M.A. Kelly, A.C. Kummel, P.C. McIntyre, Arsenic decapping and half cycle reactions during atomic layer deposition of Al_2O_3 on $In_{0.53}Ga_{0.47}As(001)$, *Appl. Phys. Lett.* **96**, 252907 (2010).

19 E.J. Kim, E. Chagarov, J. Cagnon, Y. Yuan, A.C. Kummel, P.M. Asbeck, S. Stemmer, K.C. Saraswat, P.C. McIntyre, Atomically abrupt and unpinned $Al_2O_3/In_{0.53}Ga_{0.47}As$ interfaces: experiment and simulation, *J. Appl. Phys.* **106**, 124508 (2009).
20 S. Swaminathan, Y. Oshima, M.A. Kelly, P.C. McIntyre, Oxidant prepulsing of Ge(100) prior to atomic layer deposition of Al_2O_3: *in situ* surface characterization, *Appl. Phys. Lett.* **95**, 032907 (2009).
21 E.R. Mysak, J.D. Smith, J.T, Newberg, P.A. Ashby, K.R. Wilson, H. Bluhm, Competitive reaction pathways for functionalization and volatilization in the heterogeneous oxidation of coronene thin films by hydroxyl radicals and ozone, *Phys. Chem. Chem. Phys.* **13**, 7554 (2011).
22 H. Grönbeck, Mechanism for NO_2 charging on metal supported MgO, *J. Phys. Chem. B* **110**, 11977 (2006).
23 H. Siegbahn, K. Siegbahn, ESCA applied to liquids, *J. Electron Spectrosc. Relat. Phenom.* **2**, 319 (1973).
24 H. Fellner-Feldegg, H. Siegbahn, L. Asplund, P. Kelfve, K. Siegbahn, ESCA applied to liquids IV. A wire system for ESCA measurements on liquids, *J. Electron Spectrosc. Relat. Phenom.* **7**, 421 (1975).
25 R.W. Joyner, M.W. Roberts, K. Yates, A 'high-pressure' electron spectrometer for surface studies, *Surf. Sci.* **87**, 501 (1979).
26 H.J Ruppender, M. Grunze, C.W. Kong, M. Wilmers, *In situ* X-ray photoelectron spectroscopy of surfaces at pressures up to 1 mbar, *Surf. Interf. Anal.* **15**, 245 (1990).
27 J. Pantförder, S. Pöllmann, J.F. Zhu, D. Borgmann, R. Denecke, H.-P. Steinrück, A new set-up for *in-situ* XP spectroscopy from UHV to 1 mbar, *Rev. Sci. Instrum.* **76**, 014102 (2005).
28 D.F. Ogletree, H. Bluhm, G. Lebedev, C.S. Fadley, Z. Hussain, M. Salmeron, A differentially pumped electrostatic lens system for photoemission studies in the millibar range, *Rev. Sci. Instrum.* **73**, 3872 (2002).
29 A. Knop-Gericke, E. Kleimenov, M. Havecker, R. Blume, D. Teschner, S. Zafeiratos, R. Schlögl, V.I. Bukhtiyarov, V.V. Kaichev, I.V. Prosvirin, A.I. Nizovskii, H. Bluhm, A. Barinov, P. Dudin, M. Kiskinova, High-pressure X-ray photoelectron spectroscopy: a tool to investigate heterogeneous catalytic processes, in: B.C. Gates, H. Knözinger, Eds: *Advances in Catalysis*, Vol. 52, Academic Press, Burlington, 2009, pp. 213–272.
30 M.E. Grass, P.G. Karlsson, F. Aksoy, B.W.M. Lundqvist, B.S. Mun, Z. Hussain, Z. Liu, New Ambient Pressure Photoemission Endstation at ALS Beamline 9.3.2, *Rev. Sci. Instrum.* **81**, 053106 (2010).
31 D.F. Ogletree, H. Bluhm, E.L.D. Hebenstreit, M. Salmeron, Photoelectron spectroscopy under ambient pressure and temperature conditions, *Nuclear Instr. Methods A* **601**, 151 (2009).
32 M. Salmeron, R. Schlögl, Ambient pressure photoelectron spectroscopy: a new tool for surface science and nanotechnology, *Surf. Sci. Rep.* **63**, 169 (2008).
33 H. Bluhm, M. Havecker, A. Knop-Gericke, M. Kiskinova, R. Schlögl, M. Salmeron, *In situ* photoemission studies of gas/solid interfaces at near atmospheric pressures, *MRS Bull.* **32**, 1022 (2007).
34 H. Bluhm, Photoemission spectroscopy under humid conditions, *J. Electron Spectrosc. Relat. Phenom.* **177**, 71 (2010).
35 V.E. Henrich, P.A. Cox, *The Surface Science of Metal Oxides*, Cambridge University Press, Cambridge, 1994.

36. J. Wollschläger, J. Viernow, C. Tegenkamp, D. Erdos, K. M. Schroder, H. Pfnur, Stoichiometry and morphology of MgO films grown reactively on Ag(100), *Appl. Surf. Sci.* **142**, 129 (1999).
37. J.T. Newberg, D.E. Starr, S. Porsgaard, S. Yamamoto, S. Kaya, E.R. Mysak, T. Kendelewicz, M. Salmeron, G.E. Brown, Jr., A. Nilsson, H. Bluhm, Formation of hydroxyl and water layers on MgO films studied with ambient pressure XPS, *Surf. Sci.* **605**, 89 (2011).
38. I. Lindau, P. Pianetta, S. Doniach, W. Spicer, X-ray photoemission spectroscopy, *Nature* **250**, 214 (1974).
39. K. Kobayashi, Hard X-ray photoemission spectroscopy, *Nuclear Instr. Methods A* **601**, 32 (2009).
40. M. Gorgoi, S. Svensson, F. Schäfers, G. Öhrwall, M. Mertin, P. Bressler, O. Karis, H. Siegbahn, A. Sandell, H. Rensmo, W. Doherty, C. Jung, W. Braun, W. Eberhardt, The high kinetic energy photoelectron spectroscopy facility at BESSY: progress and first results, *Nuclear Instr Methods A* **601**, 48 (2009).
41. I.A. Abrikosov, W. Olovsson, B. Johansson, Valence-band hybridization and core level shifts in random Ag–Pd alloys, *Phys. Rev. Lett.* **87**, 176403 (2001).

5
In situ spectroscopic ellipsometry (SE) for characterization of thin film growth

J. N. HILFIKER, J.A. Woollam Co., Inc., USA

Abstract: *In situ* spectroscopic ellipsometry is a versatile optical measurement technique for characterizing thin films. Ellipsometric measurements commonly determine thin film thickness, growth and etch rates, optical constants, surface and interface quality, composition, and other related material properties. This chapter covers the principles of *in situ* spectroscopic ellipsometry, considerations for *in situ* integration, and common applications.

Key words: *in situ* spectroscopic ellipsometry, thin film characterization, real-time process monitoring and control, optical monitoring, growth/etch rates.

5.1 Introduction

Thin films are commonplace in our everyday lives – from coatings on contact lenses and solar-efficient windows to cell phone components and flat-panel displays. The expansion of thin film use into a long list of technologies has placed critical demands on precise film measurements, and both research laboratories and commercial manufacturers have come to rely on optical characterization methods. These methods enable engineers to measure film thickness, optical constants and other related material properties rapidly and non-destructively. Optical characterization refers to a number of techniques that use light to study materials and thin films (samples).

Photometric measurements determine the change in reflected or transmitted light intensity caused by a sample (referred herein as 'R/T' to describe either reflectance or transmittance measurements). Standard R/T measurements are relatively simple experiments, but often inadequate to measure increasingly thin layers and complex multilayer structures found in modern thin film-based devices. Spectroscopic ellipsometry (SE) measurements consider the electric field properties (polarization) of the measured light once it is reflected or transmitted by the sample. SE is increasingly popular for thin film characterization because it is significantly more sensitive to very thin films (including sub-nanometer thickness), provides excellent precision, and yields enhanced information about the sample.

In situ SE offers the combined benefits of in-process measurement and high precision. This versatile characterization technique is routinely used

100 *In situ* characterization of thin film growth

to monitor growth or etch from a wide variety of processes. Measurements can be collected from samples in ultra-high vacuum (UHV), exposed to air, submerged within a liquid, or held in controlled gas environment. This versatility is primarily due to the non-invasive nature of optical measurements, where only the light beam travels through the process chamber. Common *in situ* SE measurements include:

- surface quality before processing;
- growth or etch rate, even with changes during processing;
- thin film optical constant variations with process conditions;
- fast development of optical constant libraries that consider variations in process conditions or material properties;
- ability to study many layers during a single process run;
- temperature and composition of compound semiconductors;
- real-time process control of thickness (end-point detection), growth rate, or material properties.

Furthermore, SE measurement speeds allow real-time measurements for all but the fastest processes. Data can be collected in a fraction of a second with hundreds of simultaneous wavelengths.

This chapter will focus on *in situ* SE measurements, analysis, and applications. Section 5.2 covers the principles of ellipsometry along with primary benefits of *in situ* SE. Section 5.3 introduces common measurements and analysis procedures. Section 5.4 discusses the considerations specific to *in situ* SE measurements; including mechanical, optical, and software integration. Section 5.5 explores a few important materials and application areas involving *in situ* SE; including compound semiconductors, optical coatings, photovoltaics, biological films, and nanotechnology. The chapter concludes with the outlook for *in situ* SE.

5.2 Principles of ellipsometry

Ellipsometry measurements describe the polarization change caused by interaction of a light beam with a sample. The polarization change itself is not generally the property of interest, but can be used to determine material properties such as thin film thickness, optical constants, and microstructure. This section introduces the fundamentals of ellipsometry measurements, basic instrumentation, data analysis procedures, and typical thin film characterization capabilities. We further consider *in situ* ellipsometry and the benefits of SE measurements during film growth or processing.

5.2.1 Spectroscopic ellipsometry (SE)

Ellipsometry measures the change in polarization that occurs when a light beam is reflected from or transmitted through a sample. For reflected light, this change in polarization, ρ, is often described with two values, Ψ and Δ (Fujiwara, 2007, p. 347):

$$\rho = \tan(\Psi)e^{-i\Delta} = \frac{\tilde{R}_p}{\tilde{R}_s} \qquad [5.1]$$

where Ψ is the ratio of reflected amplitudes and Δ is the phase difference produced upon reflection. Polarization changes arise from the reflectivity difference between electric field components oriented parallel (p-) and perpendicular (s-) to the plane of incidence, as shown in Figure 5.1. Measurements generally occur at an oblique angle of incidence (ϕ), defined as the angle between the incident light beam and sample surface normal. In Fig. 5.1, the incident light is linearly polarized with electric-field components in both p- and s-planes. In general, light reflection produces a change in amplitude ratio and phase difference between the p- and s-components. SE measures this 'change' in polarization to help determine the sample properties.

Basic ellipsometry measurements produce two quantities (typically expressed as Ψ and Δ) at each wavelength and angle of incidence. SE data are often taken at many wavelengths, leading to thousands of measured data sets. The amount of collected data can further be increased by varying the angle of incidence, but not all measured information is equally useful. Although certain wavelengths and angles are more sensitive to material properties, multiple measurements can contain equivalent information (Hilfiker et al., 2008b).

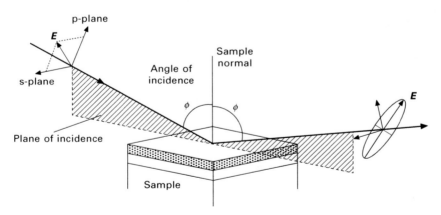

5.1 When linearly polarized incident light, consisting of p- and s- orthogonal polarization components, is reflected from a surface at oblique angle of incidence (ϕ) the result is often elliptical polarization. Ellipsometry measurements determine the change in polarization that occurs when light interacts with the sample.

To understand how optical measurements characterize thin film properties, consider the interaction of light with a single-layer coating as shown in Fig. 5.2. When light arrives at the surface, a portion is reflected and the remaining light enters the thin film. Light traveling through the film will recombine with the light reflected from the film surface, but with delayed phase and different amplitude for both p- and s-components. Thus, the light measured by SE (the recombined light) will depend on the thickness and optical constants of the film, represented by either complex refractive index (Fujiwara, 2007, p. 347):

$$\tilde{n} = n + ik \quad [5.2]$$

where n is the index of refraction and k is the extinction coefficient, or by complex dielectric function (Fujiwara, 2007, p. 347):

$$\tilde{\varepsilon} = \varepsilon_1 + i\varepsilon_2 \quad [5.3]$$

The two are related by:

$$\tilde{\varepsilon} = \tilde{n}^2 \quad [5.4]$$

Thus, SE data contain information about the surface, thin film, and substrate because the detected light includes multiple components which have traveled various paths in the sample.

SE data may also exhibit coherent interference effects from multiple light beams. For thick, transparent films the data will oscillate when plotted against wavelength because the wavelength influences whether the phase of recombined light is aligned for constructive or destructive interference. Figure 5.3 shows measured SE spectra for two transparent thin films. The thicker layer (350 nm) exhibits additional oscillations in the measured spectra

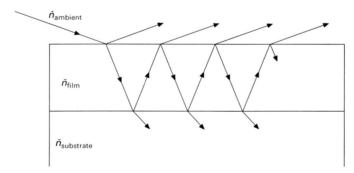

5.2 A light beam interacts with a thin film structure. At each interface, a portion of light is reflected and transmitted. The ellipsometer measures the polarization state of light from regions and interfaces in the sample. With planar interfaces and coherent light, interference can occur.

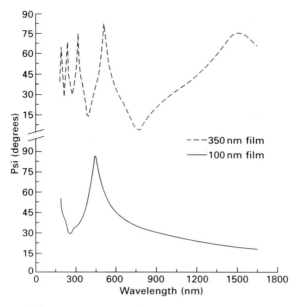

5.3 SE measurements from two transparent thin films on silicon substrate – one 350 nm and the other 100 nm thick. The number of data oscillations is a result of film thickness-dependent coherent light interference.

compared with the thinner layer (100 nm) because of the longer optical path length.

SE spectra also reveal the optical constants of the film. Index of refraction (n) describes the phase velocity for light traveling within the material and determines angle of refraction at each interface described by Snell's Law (Hecht, 1987, p. 84). Extinction coefficient (k) determines how quickly light is absorbed as it travels through a material. In addition, the optical constants determine how much light is reflected at each interface.

5.2.2 Data analysis

While SE measurements are sensitive to both optical constants and film thickness, they are not a direct measurement. In a few simplified cases, the equations can be inverted to calculate optical constants or film thickness from measured Ψ, Δ data (Azzam and Bashara, 1977, pp. 315–317). Fortunately, there are many cases where the sample properties are over-determined with hundreds or thousands of measured data points to determine a few unknown thin film properties. SE characterization is generally subject to the 'inverse problem' where the measurement can be predicted from a sample description, but sample properties cannot be directly calculated from the measured data.

Regression analysis is commonly used to determine the best-fit results considering all acquired data simultaneously. Figure 5.4 summarizes the process of ellipsometry analysis.

The first step is to collect SE data on the sample. The ellipsometer shines light with known polarization onto the sample and detects the reflected or transmitted polarization state. This measurement requires the following basic ellipsometer components: light source, polarization generator, sample, polarization analyzer, and detector (Fig. 5.5). The light source and polarization generator are often combined into the 'source head', which directs a light beam of known polarization onto the sample surface at oblique angle of incidence.

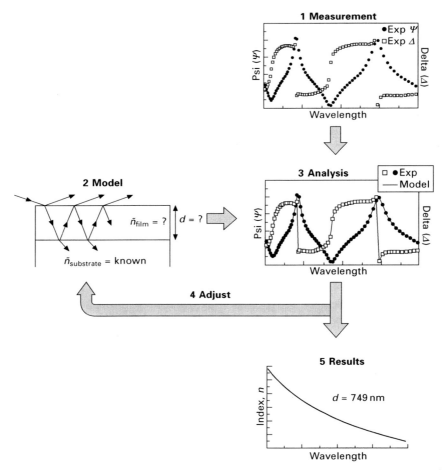

5.4 Data analysis flowchart for spectroscopic ellipsometry measurements. Regression analysis helps determine unknown parameters of the optical model such as film thickness and optical constants.

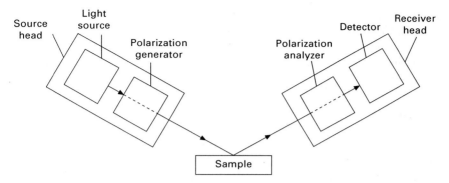

5.5 Basic ellipsometer components: light source, polarization generator, sample, polarization analyzer, and detector. In some ellipsometer designs, the source and receiver heads may combine multiple components.

The light reflects or transmits into the 'receiver head' (often consisting of both polarization analyzer and detector), which determines the new polarization state. In addition to these basic components, a spectroscopic ellipsometer must be equipped to scan and detect at various wavelengths. Many different SE designs are available, with details available in the literature (Johs *et al.*, 1999; Collins, *et al.*, 2005a, 2005b; Jellison and Modine, 2005; Fujiwara, 2007).

In step 2, the sample is described using a planar optical model, which represents layer thicknesses and optical constants for each material. The model-generated response is calculated based on Fresnel equations (Hecht, 1987, pp. 94–96). For unknown sample properties, initial estimates are also given at this step. Reference optical constant values for many materials are available in the literature (Palik, 1998a, 1998b, 1998c; Adachi, 1999). However, SE measurements are often sensitive to optical constant values in the third or fourth decimal place, so any deviation between published values and the measured material will produce errors in the model calculations. For this reason, reference optical constants often serve only as initial estimates; the model optical constants are then allowed to vary to 'best match' the measured Ψ, Δ data.

In Step 3, regression analysis is used. The unknown model parameters are varied to improve the match between model-generated curves and collected data by minimizing differences between the curves, as described by a comparator function, such as mean squared error (MSE). The model can be modified, as shown in Step 4 of Fig. 5.4, in a further attempt to improve the match between model predicted data and measurement. As a general rule, the best sample description is often the one produced by the simplest model that matches data adequately; still, it is important to ensure a unique result for each unknown sample property.

The best model provides information about the unknown sample properties, as shown in Step 5. SE measurements are commonly used to determine thin film thickness and optical constants. However, many additional material properties can be determined through their effect on material optical constants: crystallinity, conductivity, composition, porosity, strain, and surface roughness. Figure 5.6 shows the optical constants for a series of $Si_{1-x}Ge_x$ thin films with varying composition (x). As the composition changes, so do the critical points associated with electronic transition energies, which in turn causes a shift in $Si_{1-x}Ge_x$ optical constants. The composition can be deduced from SE by measuring this variability in the optical constants of a $Si_{1-x}Ge_x$ layer. For additional information regarding ellipsometry analysis procedures, see Fujiwara (2007), Jellison (2005), and Tompkins and McGahan (1999).

5.2.3 In situ SE

In addition to the fundamental advantages of SE (non-destructive, non-invasive, precise, accurate, and versatile), *in situ* SE has the ability to monitor a process of thin film deposition or etch as it occurs. Modern SE systems can perform precise measurements within a second or less, so most processes can be monitored in real time. In addition, *in situ* SE measurements can be made directly on the sample of interest, rather than a witness sample or at a separate location within a process chamber.

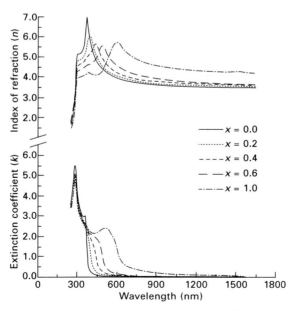

5.6 Optical constants for $Si_{1-x}Ge_x$ with different compositions, x.

In situ spectroscopic ellipsometry (SE)   107

Because modern SE optical heads have a small footprint, they can be conveniently integrated with optical ports in large chambers commonly used for deposition and etch of thin films. Figure 5.7 shows an atomic layer deposition (ALD) chamber incorporating *in situ* SE as a real-time monitor for thin film growth. For processes that require only small space, such as liquid flow-cells or variable-temperature stages, the process itself may be integrated directly into standard *ex situ* SE hardware. Figure 5.8 shows a 100 μL liquid flow-cell mounted on a conventional SE system. This cell combines *in situ* SE with a quartz crystal microbalance with dissipation monitoring (QCM-D) which allows real-time study of surface adsorption from a liquid solution.

Figure 5.9 shows an *in situ* SE measurement during growth of a dielectric film on silicon. For the purpose of illustration, data for only four wavelengths are plotted against time; actually, data at hundreds of wavelengths are collected for each measurement time. *In situ* SE can collect data during various stages

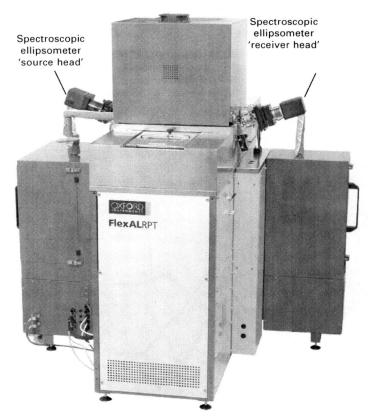

5.7 Spectroscopic ellipsometer mounted to an atomic layer deposition (ALD) process chamber (photo courtesy of Oxford Instruments).

108 *In situ* characterization of thin film growth

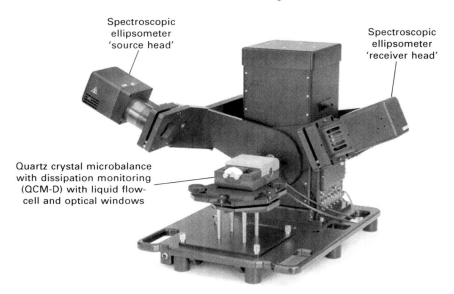

5.8 Quartz crystal microbalance with dissipation monitoring (QCM-D) with liquid flow-cell and optical windows mounted to a spectroscopic ellipsometer for combined *in situ* SE and QCM-D characterization (photo courtesy of Biolin Scientific/Q-Sense).

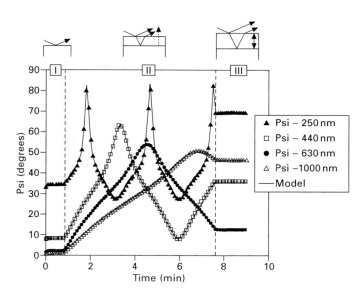

5.9 In situ SE measurements during growth of a transparent layer. Three designated regions provide information (I) before, (II) during, and (III) after processing.

of processing, charting unknown sample properties as they emerge. The substrate surface is first monitored before the film is deposited (Region I). Measurements before processing provide a 'baseline' of substrate optical constants, surface quality and measurement angle of incidence. This can be critical information for sensitive measurements, such as protein adsorption on the surface of only a few nanometers thick. For the silicon substrate in Fig. 5.9, the pre-process region is used to determine native oxide thickness and angle of incidence, two values that are not always well known *a priori*.

As a film grows, each dynamic measurement point captures the interaction with a different thickness and data continue to change with time (Region II). If *ex situ* measurements were used to collect the same information, hundreds of samples with different film thickness would be required. By contrast, *in situ* SE measurements do not interrupt growth, and provide continuously evolving information about growth rate, film thickness, and optical constants.

In Region III, the process is complete and the sample structure is measured to determine final film thickness and optical constants. This region also shows the sample before aging, or in the case of ultra-high vacuum (UHV) processes, before surface oxidation. In-line SE typically measures region I or III, but the real power of *in situ* SE is the wealth of information made available *during* the process.

5.3 *In situ* spectroscopic ellipsometry (SE) characterization

In situ SE measurements are sensitive to a wide variety of material properties and process conditions. They can also yield the dynamics of these properties during the process. This section reviews common *in situ* SE measurements, as well as basic analysis procedures that enable *in situ* SE characterization.

5.3.1 Growth and etch rate

When *in situ* SE measurements are collected during thin film growth or etch, the film thickness changes from measurement to measurement. The unknown sample properties are determined at each measurement timepoint, when adequate information is available. In many cases, to simplify data analysis, a constant growth/etch rate is assumed over a specific time period. Data analysis is further simplified if the layer optical constants are stable and do not vary with thickness. Figure 5.10 shows film thickness measured during sputter deposition of a dielectric layer on metal. Dashed lines indicate the beginning and end of deposition. At 13 minutes, sputtering power was doubled to increase film growth rate. In this experiment, thickness was determined separately for each measured time-point and growth rate calculated for each interval.

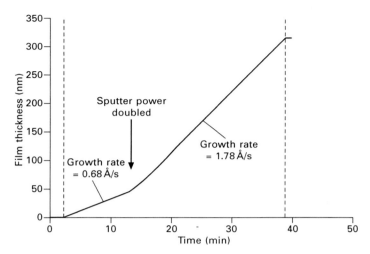

5.10 Film thickness of a dielectric layer measured by *in situ* SE during sputter deposition. The growth rate increases at 13 minutes, as the sputter power is doubled.

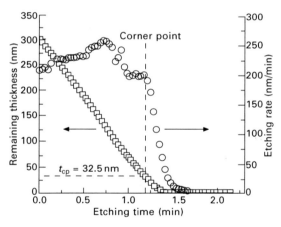

5.11 Film thickness and etch rate for a PMMA thin film during plasma etch process. At the corner point, which occurred with 32.5 nm remaining film, the etch rate decreases significantly (reprinted from Kokkoris *et al.*, *Plasma Processes and Polymers*, vol. 5, pp. 825–833, 2008 with permission from Wiley-VCH).

As another example, Kokkoris *et al.* (2008) applied *in situ* SE during plasma etch. Figure 5.11 shows film thickness and etch rate for a poly(methacrylate) (PMMA) thin film. They found the etch rate decrease as the PMMA layer reaches a thickness equal to the surface roughness. The 'corner point', where etch rate starts to decrease, allowed the authors to estimate surface roughness, which was confirmed by atomic force microscopy (AFM) measurements of the same surfaces.

Maynard et al. (1998) demonstrated end-point detection with in situ SE during plasma etch of patterned thin films. They etched optically absorbing layers such as TiN and poly-Si, which limited thickness determination to thin layers where light can penetrate the film. Metals absorb light at wavelengths across most of the typical spectrum, which prevents thickness measurements for thick layers (typically > 50 nm). Thin layers can be measured, but require special techniques to determine both optical constants and thickness (Hilfiker et al., 2008a).

An experiment by Pribil et al. (2004) combined in situ transmission SE and transmitted intensity measurements to determine the optical constants for thin metal layers deposited on glass with magnetron sputtering up to a thickness of 35 nm. The additional information available from simultaneous SE and T proved essential as metal optical constants did vary with thickness. Figure 5.12 shows the evolution of thickness-dependent optical constants for cobalt and titanium films (Pribil et al. 2004).

Wide spectral ranges available with modern SE systems help expand characterization for materials which remain semi-transparent in a portion of the measured spectrum. Consider the in situ SE measurements in Fig. 5.13 for deposition of an AlGaAs thin film on GaAs. Data at three wavelengths are graphed versus time to demonstrate the effect of film absorption on in situ SE data. Data oscillations correspond to increasing film thickness, where coherent interference still occurs between the surface reflection and light traveling through the thin film. The oscillations are quickly dampened at 350 nm, where the AlGaAs film is strongly absorbing. Data at both 475 nm and 600 nm continue to oscillate throughout the growth because at these wavelengths the AlGaAs remains semi-transparent.

5.3.2 Virtual interface approaches

Standard SE analysis requires the model response is calculated from every layer in the stack. Errors from underlying films propagate through the entire structure. A virtual interface (VI) can approximate the underlying structure as a single interface, rather than tracking the entire sample history. The VI is placed towards the surface, as shown in Fig. 5.14. Here the VI intersects the third layer and growth is modeled on this interface with no knowledge retained for the underlying structure. There are various methods for calculating a VI (Aspnes, 1993, 1996; Urban and Tabet, 1993; Kouznetsov et al., 2002; Johs, 2004), but the mathematical details are beyond the scope of this chapter.

The common pseudo-substrate approximation (CPA) described by Aspnes (1993, 1996) converts the measurement at this point into a 'pseudo-substrate'. The main shortcoming of the CPA is that it requires underlying materials

112   *In situ* characterization of thin film growth

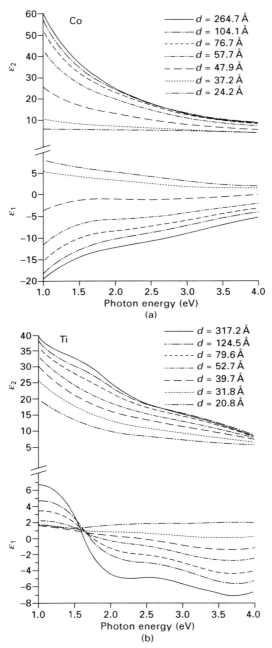

5.12 Optical constants for thin metal films determined from combined *in situ* SE and transmission intensity measurements. Optical properties for (a) cobalt and (b) titanium thin films change with thickness within the measured range up to 265 Å and 317 Å, respectively (adapted from Pribil *et al.*, 2004, with permission).

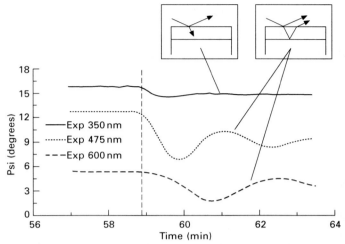

5.13 In situ SE measurement of AlGaAs film deposition shows oscillations due to coherent interference when light can travel through the film and return to the surface. Light at 350 nm is strongly absorbed by the film, resulting in dampened interference oscillations. Data at 475 nm and 600 nm continue to oscillate as thickness increases, due to lower absorption at these wavelengths.

5.14 Virtual interface approach can simplify *in situ* SE data analysis by approximating the underlying structure as a single interface.

with either high refractive indices or strong absorption. Thus, it fails with transparent layers common to optical coating stacks. Other VI approaches have been proposed using multiple time-points to allow a direct solution. Urban and Tabet (1993) offer a calculation involving five time-points to balance the number of unknowns and measured values. More recently, Johs (2004) proposed a general virtual interface (GVI) algorithm applicable with any type of material using a minimum of three time-points. Johs demonstrated the GVI approach during deposition of a diamond-like carbon film on metal in a plasma-enhanced chemical vapor deposition (PECVD) chamber.

5.3.3 Multilayer and graded film characterization

Multilayers, with their many unknown properties, are often too complex for *ex situ* characterization. *In situ* SE provides two key advantages: first, each layer of the structure can be studied individually, and second, multilayer modeling can often be simplified with a VI approach. Graded films are modeled in a similar manner to multilayers, with a large number of layers with varying optical constants to approximate the film variation. For both situations, a VI avoids the complex thin film stack calculation for the entire sample structure.

Kim and Collins (1995) demonstrated *in situ* SE analysis during PECVD of compositionally graded amorphous silicon–carbon alloys. The H_2-dilution ratio was adjusted to produce void fraction variations through the film, which are measured by their effects on optical constants using a CPA approach. Amassian *et al.* (2002) used GVI analysis to determine the refractive index profile (Fig. 5.15) of 'an apodized rugate filter deposited by PECVD from a continuously varying $TiCl_4/SiCl_4$ mixtures', which produces mixed TiO_2–SiO_2 films.

5.3.4 Process control

Because *in situ* SE measurements can be collected in a fraction of a second, they offer an opportunity to implement real-time process control, which is important in many industrial applications. Real-time ellipsometry control was first demonstrated by Aspnes *et al.* (1992) for a parabolic quantum well structure. In this study, a single wavelength was used to determine composition of the outermost surface during AlGaAs deposition with feedback

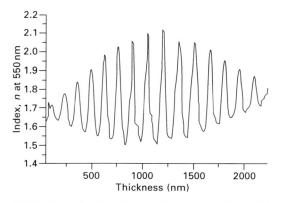

5.15 Index of refraction profile versus film thickness measured by *in situ* SE for an apodized rugate filter (reprinted from Amassian *et al.*, 45th Annual Technical Conference Proceedings-SVC, pp. 250–255, 2002, with permission from Society of Vacuum Coaters).

to the aluminum source for composition control. Though the composition profile was controlled, well-width was established assuming growth rates from previous calibration.

Multi-wavelength ellipsometry expanded the ability to control processes in real-time. Growth-rate control of CdTe from metal-organic vapor phase epitaxy (MOVPE) was demonstrated using a 12-wavelength ellipsometer (Johs et al., 1993). The feedback of voltage to the Cd mass flow controller was limited to every 3 seconds due to data acquisition and computer speed at the time. Kildemo, et al. (1996) used a three-wavelength in situ ellipsometry measurement to control PECVD deposition of optical coating multilayers with 1% accuracy. Because the coatings were transparent, a standard VI approach was not successful. Instead, they predicted data trajectories in advance of deposition for the measured structure. During the process, the layer deposition was stopped when real-time experimental data approached the calculated end-points.

Herzinger et al. (1996) demonstrated multilayer growth control of an 8-period Bragg reflector structure with a VI approach that considers multiple measurement times to adjust the model interface. Although analysis on a 90 MHz Pentium computer required 3 to 8 seconds per measurement time-point, precise shutter control was achieved through stop-point predictions based on the most recent measured thicknesses, as shown in Fig. 5.16. Successful control to 1.3% thickness accuracy was achieved using a 44-wavelength ellipsometer under non-ideal conditions (beam wobble, window birefringence, and non-optimum angle) with an intentionally drifting process. Today, modern computers and advanced SE systems with hundreds or even thousands of wavelengths make in situ SE control significantly better and easier.

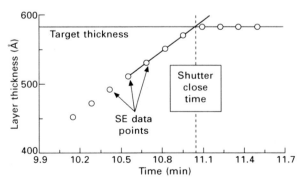

5.16 Demonstration of MBE growth control for a 583 Å GaAs layer (adapted from Herzinger et al., 1996, with permission from the Materials Research Society).

5.3.5 Process-dependent optical constant libraries

Material properties such as crystallinity, composition, conductivity, and strain affect the optical constants of many materials. In addition, process conditions such as temperature, pressure, and growth rate can also affect optical constants. Because *in situ* SE can precisely measure optical constants, this technology can also be used to monitor material and process conditions.

Volintiru, *et al.* (2008a) studied the effects of process conditions on the refractive index of aluminum oxide thin films. They observed that as substrate temperature increased during remote plasma-enhanced metal-organic chemical vapor deposition (MOCVD), film density also increased which resulted in higher refractive index. They also observed further index increases with increasing bias voltages, as shown in Fig. 5.17.

Where the same process is applied to the same material (e.g. commercial production), *in situ* SE is often used to create process-dependent or material-dependent optical libraries. A series of materials with varying properties are measured to determine optical constants related to each condition. A critical point shifting algorithm (Snyder *et al.*, 1990) is widely used to build material libraries which vary the optical constants versus material property or process condition. This approach can also be applied to materials with two varying properties, such as quaternary compound semiconductors or ternary semiconductors with both composition and temperature variation. Figure 5.18 shows the complex dielectric function for a series of $Hg_{1-x}Cd_xTe$ thin films with varied composition, x, determined *in situ* at growth temperature of 180 °C.

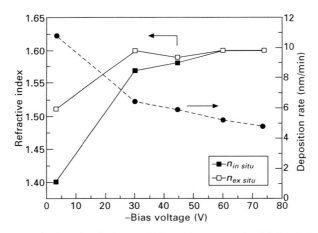

5.17 Refractive index at 633 nm (measured with both *in situ* SE and *ex situ* SE) and deposition rate versus bias voltage for aluminum oxide thin films measured by *in situ* SE (reprinted from Volintiru *et al.*, 2008a, with permission from Wiley-VCH).

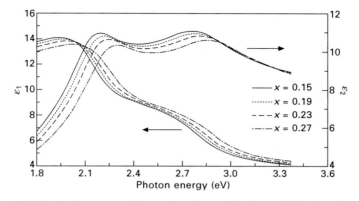

5.18 Optical constant library developed for $Hg_{1-x}Cd_xTe$ thin films at 180 °C.

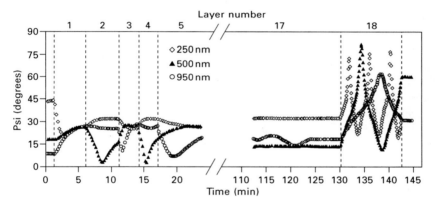

5.19 In situ SE data from a sputter-process run with 18 various Ta, a-Si, and SiO_x layers. Process conditions were varied for each layer to calibrate their effect on film properties.

Material libraries can be created from a series of samples measured *ex situ*. However, this is a tedious and slow process, often with sample exposure to film-changing environments like oxygen or moisture. The real benefits of *in situ* SE are the ability to measure (i) samples under varying process conditions not available *ex situ*, (ii) virgin surfaces with little to no oxide, (iii) surfaces exposed to controlled atmosphere, and (iv) a large number of thin films within a single process run. For the latter, virtual interface modeling can help determine film properties without concern for the underlying sample structure. Figure 5.19 shows *in situ* SE data from a complex process run with a total of 18 different thin films. Process conditions, such as sputter gun current and oxygen flow, were varied for each layer to produce series of different Ta, a-Si, and SiO_x layers.

Data analysis of each layer is facilitated with a virtual interface to avoid, in some cases, tens of under-layers in the model. From this single run, growth rates for both Ta and a-Si films were calibrated, as shown in Fig. 5.20. In addition, the evolution of optical constants for each material was determined in relation to process condition.

5.3.6 Surface and interface quality

The surface sensitivity of SE measurements encourages study of both interface and surface quality. Surface quality can be an important measurement before and after growth. *In situ* SE can also monitor the surface development during processing. Johs *et al.* (1997) demonstrated *in situ* SE monitoring of a CdZnTe substrate surface before $Hg_{1-x}Cd_xTe$ film deposition, as shown in Fig. 5.21. SE measurements determined both surface quality and substrate temperature during the heat-clean procedure. As temperature increased, amorphous Te desorbed from the surface, followed by oxide desorption to produce a clean underlying surface for film growth.

Surface evolution can also be studied during growth. Fujiwara *et al.* (2003) monitored the formation of interface and surface roughness during microcrystalline silicon film growth. Figure 5.22 shows the surface, interface, and bulk layer thickness evolution for two microcrystalline silicon films on different substrates. An interface of 18 Å forms on the ZnO:Ga surface, but not on SiO_2, which was confirmed with cross-sectional transmission electron microscopy (TEM) on the same samples. Evolution of bulk film and surface are similar for both samples, after interface formation is complete.

In situ SE measurements during the process are sensitive to interface development. Johs *et al.* (1998a) compared two interfaces for film growth

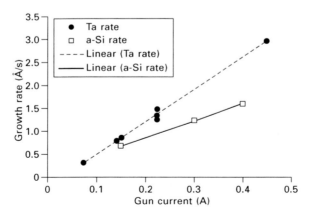

5.20 Ta and a-Si growth rate calibration versus gun current developed from *in situ* SE measurements of multiple layers.

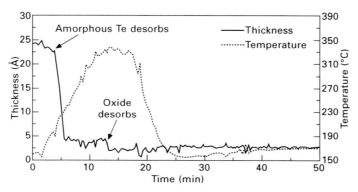

5.21 In situ SE during the preparation of CdZnTe substrate before HgCdTe film growth provides surface layer thickness and substrate temperature. As the substrate is heated, an amorphous Te layer is desorbed, followed by the native oxide desorbtion – ultimately producing an ideal surface to start film growth (reprinted from Johs et al., III-Vs Review, vol. 10, no. 5, 'Real-time process control with in situ spectroscopic ellipsometry', pp. 42, 1997, with permission from Elsevier).

on InP. The $In_{0.53}Ga_{0.47}As$ layer produced an ideal interface, while the poor data fit quality for $In_{0.52}Al_{0.48}As$ layer suggested a rough interface. This was substantiated by successfully modeling the rough interface with an effective medium approximation (EMA).

5.4 *In situ* considerations

In this section, we review basic integration issues specific to *in situ* SE measurements, including:

- mechanical integration;
- sample alignment;
- rotating samples with beam wobble;
- window concerns;
- environmental conditions;
- software integration.

5.4.1 Mechanical integration

One of the most important requirements for *in situ* ellipsometry is optical access to the sample during processing. The ellipsometer source and receiver 'heads' are typically external to the process area, so multiple holes are needed in the chamber at appropriate locations for incoming and outgoing light beams. A clear path for the light beam within the process chamber is

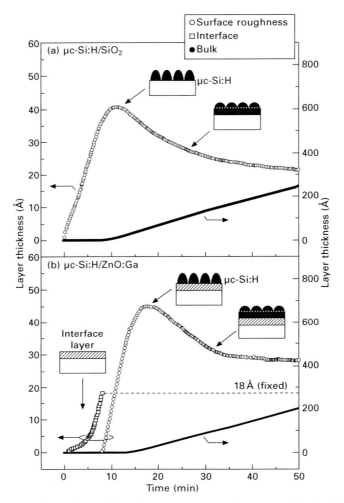

5.22 Evolution of interface, surface roughness, and bulk thickness measured by *in situ* SE during growth of microcrystalline silicon films (reprinted from Fujiwara *et al.*, 2003, with permission from the American Institute of Physics).

also necessary. Optical ports should be positioned to allow an appropriate angle of incidence between the measurement light beam and sample surface. A common limitation of *in situ* measurements is a fixed angle of incidence, so angle choice can be important. Preferred angles are between 60° and 75° relative to the sample normal. However, angles from 35° to 85° may be acceptable, although with reduced sensitivity, depending on the sample to be studied.

There are three considerations for appropriate *in situ* angles. First, is adequate optical access to the process chamber available for *in situ* SE? When necessary, beam steering via prisms or mirrors may be necessary to manipulate the measurement beam to an appropriate angle of incidence. Second, does the angle provide measurement sensitivity to the material properties of interest? This is less critical with modern ellipsometers, which collect accurate SE measurements even at non-optimal angles. However, the polarization change is reduced as angle of incidence moves significantly away from the Brewster condition and angles below 45° are rarely useful. Third, does the angle produce a large spot size on the sample? As angle of incidence increases, so does the projected length of the light beam on the substrate. This can be important for small or laterally non-uniform samples. The spot size on the sample surface will remain the width of the beam diameter, but with a projected length along the plane of incidence dependent on angle of incidence (ϕ) as:

$$\text{Length}_{\text{spot}} = \frac{\text{Diameter}_{\text{beam}}}{\cos(\phi)} \quad [5.5]$$

Thus, the spot length projects to 2×, 3×, and 4× the beam diameter near 60°, 70°, and 75°, respectively.

Figure 5.23 shows the integration of an *in situ* SE on sputter deposition chamber (Woollam, *et al.*, 1996). An existing chamber was retrofit with ports positioned at 75° to monitor above one of four sputter guns. In the molecular beam epitaxy (MBE) system used by Herzinger *et al.* (1996), *in situ* SE ports were adapted to existing ports for reflection high-energy

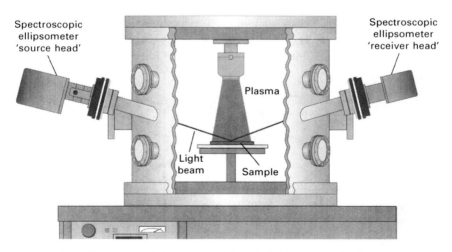

5.23 Mechanical integration of *in situ* SE to Kurt J. Lesker sputter deposition chamber (adapted from Woollam *et al.*, *SPIE Proc.*, vol. 2873, pp. 140–143, 1996, with permission from SPIE).

electron diffraction (RHEED). This forced a non-ideal angle of incidence above 80°, but avoided additional chamber modifications to allow access for SE. A general-purpose angle of incidence is 70° and many commercial process chambers are now offered with existing ports at this angle to allow *in situ* SE.

Modern SE 'heads' are compact and allow convenient attachment to most chambers. However, space on a busy chamber may still be difficult to find. Beam steering has been implemented, with the use of prisms or mirrors, to manipulate the light beam in and out of a chamber. Figure 5.24 shows a special 'prism-assisted' method to allow optical access when optical ports are not available, through a chamber lid positioned parallel to the sample surface (Johs, 1999). Special care is required with extra reflections to calibrate the polarization effects, track possible changes to the plane of incidence, and determine the actual angle of incidence at the sample surface. Fiber optics also occupy a role in moving light from one point to another for optical measurements. However, fiber optics do not maintain light polarization over a wide spectral range, so SE polarizing components remain between the fiber and sample.

The source head is positioned to send the incoming light beam to the sample. Some tip-tilt adjustment may be necessary. The receiver head is positioned to collect the measurement beam after interaction with the sample. This may require translation of the receiver position; depending on chamber geometry, sample angle precision, and measurement beam size. Large chamber dimensions exacerbate the need for receiver translation. Tip-tilt adjustment of the receiver head may also be required to align the collected beam within the receiver optical path.

Small process chambers (e.g. liquid and temperature cells) may be mounted to an *ex situ* SE system. Here, the alignment steps typically involve movement (tip–tilt–translation) of the process chamber relative to the SE system. The

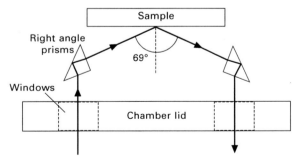

5.24 Mechanical integration with 'prism-assisted' access to a vacuum chamber through a chamber lid parallel to the sample surface (adapted from Johs, 1999).

process chamber is often aligned such that incoming light beam enters normal to the entrance window. A well-machined design can mitigate many of the alignment issues.

With proper alignment, the *in situ* SE will capture the reflected light beam from the sample. However, SE is typically sensitive to angle of incidence to ±0.01° and thus requires accurate knowledge of correct angle to maintain high-accuracy measurements. Angle of incidence can be altered by alignment and should be determined prior to measurement. This is commonly achieved by measuring a known sample, such as SiO_2 on Si. The optical constants for both crystalline Si and thermal SiO_2 are well-known (Herzinger *et al.*, 1998), which allows ellipsometric characterization of the SiO_2 thickness and angle of incidence. The measured angle will remain accurate provided the sample position is retained. Alternately, the angle can often be determined during *in situ* measurements, avoiding the need to determine angle from a standard sample.

The SE alignment depends on the reproducibility of sample position. If the sample is displaced along the sample normal (d_z) as shown in Fig. 5.25(a), it will lead to an offset of the reflected measurement beam at the receiver ($d_{receiver}$), according to:

$$d_{receiver} = 2d_z \sin(\phi) \qquad [5.6]$$

where ϕ is the angle of incidence. Thus, the beam displacement at the receiver

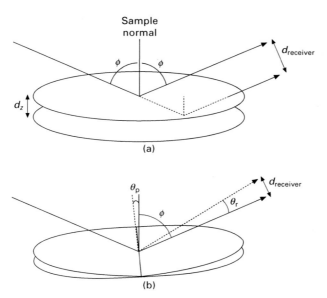

5.25 The reflected beam position at the receiver plane is displaced due to (a) sample offset normal to the surface or (b) sample tilt.

will equal the sample displacement at 30° angle of incidence and exceed 1.8× the sample displacement for angles above 65°.

If the sample is tilted, the resulting beam displacement depends on tilt direction relative to the plane of incidence. Figure 5.25(b) shows a sample tilted in the plane of incidence (θ_p), which deviates the reflected beam angle (θ_r) by:

$$\theta_r = 2\theta_p \qquad [5.7]$$

The angular beam deviation for sample tilt perpendicular to the plane of incidence (θ_s) can be calculated from (Liphardt, 2011):

$$\cos(\theta_r) = \sin^2(\phi) + \cos^2(\phi)\cos(2\theta_s) \qquad [5.8]$$

This produces an offset beam position at the receiver ($d_{receiver}$), dependent on the distance (D) between the sample and receiver, as:

$$d_{receiver} = D\tan(\theta_r) \qquad [5.9]$$

Small tilt angles in the p- and s-planes produce beam displacement in the p- and s-planes, respectively. At 75° angle of incidence, a sample tilt of 1° will deviate the reflected beam by 0.5° to 2.0°, depending on tilt direction relative to the plane of incidence. This can produce significant offsets in the reflected beam position at the receiver, particularly with large chamber dimensions. For 'D' equal to 500 mm, a 1° tilt in p- and s-planes would offset the beam location by 17.5 mm and 4.5 mm, respectively.

Many SE systems integrate alignment detectors in the receiver head to allow alignment to the measurement beam. These detectors can also be used to track alignment changes from one sample to the next, or even during a process. This becomes increasingly important when considering beam wobble due to substrate rotation, which is covered in the next section.

5.4.2 Sample rotation and beam wobble

To maintain film uniformity, it is common to rotate the sample during processing. However, any misalignment between sample surface normal and the rotation axis will produce wobble and affect the reflected beam position, as shown in Fig. 5.26. The sample wobble produces a nearly ellipsoidal beam trajectory at the receiver plane and affects the overall accuracy of *in situ* SE measurements. The length and width of the ellipsoidal beam trajectory can be calculated based on Equations [5.7]–[5.9].

$$\text{Length} = 2D\tan(2\theta) \qquad [5.10a]$$

$$\text{Width} = 2D\tan\{\cos^{-1}[\sin^2(\phi) + \cos^2(\phi)\cos(2\theta)]\} \qquad [5.10b]$$

Further, for small tilt angles, the width is approximately:

$$\text{Width} \approx 2D\tan(2\theta)\cos(\phi) \qquad [5.10c]$$

In situ spectroscopic ellipsometry (SE)

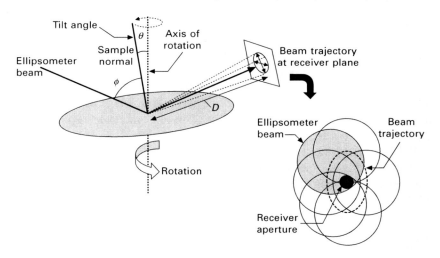

5.26 Substrate rotation can produce beam wobble when the sample normal is not aligned with the axis of rotation. This produces an elliptical trajectory of the reflected beam at the receiver plane. A large beam diameter can maintain overlap with the receiver aperture as the beam precesses (adapted from Johs *et al.*, *Thin Solid Films*, vol. 313–314, pp. 490–495, 1998b, with permission).

Maximum beam displacement is generally of most concern, and will occur in the plane of incidence, according to Equation [5.10a].

Preferred levels of beam wobble are less than ±0.1°, but this can be difficult to maintain, even with excellent mechanical design. Johs *et al.* (1998b) demonstrated that wobble as high as ±0.5° can be mitigated with the following methods: detector overfill, data synchronization, special alignment and calibration practices, and zone-averaging.

It is critical that the reflected light reach the detector, but this can be difficult to maintain and depends on chamber dimensions and the amount of wobble. A beam with a large diameter will overfill the detector aperture and thus ensure that signal reaches the SE receiver, as shown in Fig. 5.26. In this scenario, only a portion of the light beam is collected, which reduces the signal-to-noise ratio. Fortunately, the measurements are still accurate because SE data are a ratio of two polarization directions.

Both angle of incidence and plane of incidence can change as the beam precesses. If measurement time is a fraction of the sample rotation speed, each measurement will collect a different time-averaged range of beam locations. To improve data accuracy, Johs *et al.* (1998b) suggest synchronizing the measurement averaging time and sample rotation speed. This eliminates 'beating' effects in the data, caused by consecutive measurements that collect different portions of the beam precession.

It is also important to align the sample and calibrate the ellipsometer while the sample is rotating under measurement conditions. This optimizes SE alignment, calibration, and calculation of the average angle of incidence. For high-accuracy measurements, zone-averaging is also recommended as a means to cancel the first-order ellipsometry errors due to misaligned plane of incidence. Zone-averaging requires multiple SE measurements with polarizing optics at positive and negative orientations. This does increase measurement time which may not be feasible for rapid processes.

An alternate method to mitigate beam wobble has been demonstrated by both Haberland *et al.* (1998) and Johs and He (2011). A spherical mirror which reflects the beam back onto the sample is attached opposite the source head. The receiver head is located near the source head, collecting the return-beam. This 'return path' design can cancel beam precession at the receiver, which improves signal-to-noise. Haberland *et al.* (1998) concluded that this method can easily compensate for wobble of nearly 2°. The optical model used with such 'return-path' SE must consider that the measurement beam reflects from the sample twice. In addition, any mirrors or prisms that direct the light beam to the receiver head can produce polarization effects, which also need to be considered. Synchronization of the measurement time and rotation speed is beneficial, as the angle of incidence and plane of incidence will still vary on the sample even as the beam wobble is mitigated. Figure 5.27 shows one realization of this method using a prism to direct the source-beam into the chamber using the same window as the reflected beam traveling back to the receiver (Johs and He, 2011). Under standard conditions, tests using ±0.8° substrate wobble and 1 m distance between sample and receiver produced about 28 mm beam precession at the receiver. Implementing the 'return-path' approach reduced this precession to less than 1 mm.

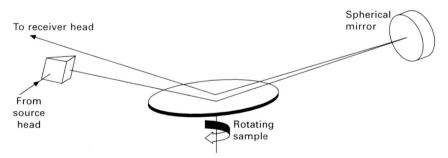

5.27 'Return-path' ellipsometry method to compensate for substrate wobble uses a spherical mirror to reflect light back on the sample. Both source and receiver heads use the entrance window, with a prism to direct the source beam in this implementation (adapted from Johs and He, 2011).

5.4.3 Window concerns

Windows often separate *in situ* SE optics from the process. However, windows can affect the light beam through birefringence, reduced intensity, and possible beam deviation or displacement. Window birefringence will modify the light beam polarization and affect SE measurements; because of this, anisotropic window materials, such as sapphire, are avoided. Amorphous materials, such as fused silica glass, can have strain-induced birefringence of a couple of degrees. Fortunately, calibration methods can successfully correct the small window birefringence effects of common vacuum windows (Jellison, 1999; Johs, 2000; Johs *et al.*, 2001). Strain-induced birefringence often depends on window mounting, so calibration is recommended when windows are repositioned, remounted, placed under vacuum, or modified by cleaning or heating.

Window retardance effects depend not only on the window material but also on its dimensions. Standard fused silica flats can have 3° to 5° retardance, which is manageable through proper calibration. For very sensitive applications, 'strain-free' windows, such as those from BOMCO Inc., exhibit only 0.1° to 0.5° retardance. Special care should be taken when studying anisotropic samples, as traditional window calibrations are designed for isotropic measurements and may not correctly anticipate cross-polarization possible from anisotropic samples.

Another undesirable condition occurs if windows become coated with an absorbing material due to their close proximity to the deposition process. This will affect signal-to-noise, but not SE accuracy as both p- and s-polarizations are equally absorbed. Window coating can be a more significant problem for *in situ* R/T, which measure light intensity, not polarization. To reduce window coating, optical ports can be placed as far from the process as possible, gated, heated, or even purged to reduce window exposure. Eventually, windows may need to be cleaned or replaced.

Windows can also deviate the beam direction if the light beam is incident non-normal to a window or the window is wedged. Figure 5.28 shows an offset in beam position, δ_{window}, for a planar window, which can be calculated for a given window thickness, d_w, as:

$$\delta_{\text{window}} = \frac{\sin(\phi_1 - \phi_2) d_w}{\cos\phi_2} \quad [5.11]$$

where the refracted angles (ϕ_2, ϕ_3) can be calculated from the incident angle (ϕ_1) using Snell's Law (Hecht, 1987, p. 84):

$$n_1 \sin\phi_1 = n_2 \sin\phi_2 = n_3 \sin\phi_3 \quad [5.12]$$

This effect is generally trivial unless the ambient index (n_1, n_3) differs on each side of the window, as is common for liquid cells, where the window

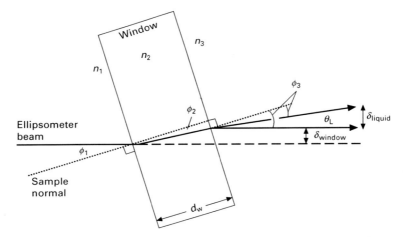

5.28 Light beam is offset (δ_{window}) through a window when entering at non-normal incidence. If ambient index is different on each side of the window (e.g. liquid cell), an angular deviation (θ_L) can also occur.

is in contact with air on one side, and with the liquid on the other. The light beam is redirected with both the displacement, δ_{window}, and an angular deviation, θ_L, as shown in Fig. 5.28. Final beam location can be determined by adding window displacement of Equation [5.11] with further displacement through the liquid, calculated as:

$$\delta_{liquid} = \tan(\theta_L)D = \tan(\phi_1 - \phi_3)D \qquad [5.13]$$

for a path length through the process (D). Of course, this is only an estimate, as beam deviation may also occur at the exit window. The key to reduce beam deviation is by either aligning the measurement beam normal to the entry window or reducing path length within the liquid environment (Hilfiker, 1995).

With a short path length, beam deviation may not be a concern. Byrne *et al.* (2009) demonstrated *in situ* SE with a tubular flow cell that allows variable angle of incidence measurements. While the light beam does not enter a flat window, the short path length within the cell minimizes beam deviation. Their tubular design provides the additional capacity to map the sample along a linear direction, a feature that was utilized to study protein adsorption on binary graded metals, such as $Al_{1-x}Ta_x$.

5.4.4 Process condition

SE measurements are compatible with most process conditions. However, background light or ambient absorption can have an effect on certain measurements. For example, some plasma processes produce bright light, which may affect the DC component of the SE signal. Fortunately, most modern SE designs are chopped or optically modulated by rotation or modulation of the polarizing optics. Using the AC signal components isolates the SE measurement from fluctuations in ambient light level.

Since SE is also a non-invasive measurement, it is not generally affected by process environment. Light absorption is not typically a problem in vacuum or air, although oxygen and/or water vapor will strongly absorb vacuum ultraviolet (VUV) wavelengths. Thus, VUV ellipsometry requires vacuum or purge throughout the entire optical path (Hilfiker, 2005). Various gases, such as CO_2, also absorb some bands in the infrared spectrum.

Light absorption in liquid ambient is more common. For this case, cell designs reduce path length within the liquid or avoid traveling within the liquid altogether. A standard liquid cell design is shown in Fig. 5.29(a). Here, light travels through the liquid to reach the sample surface. Zudans, *et al.* (2003) demonstrated *in situ* liquid interface studies with a measurement beam entering the backside of a glass substrate, as shown in Fig. 5.29(b). Poksinski and Arwin (2004) used a prism to accommodate internal reflection ellipsometry, which also bypasses the liquid, as shown in Fig. 5.29(c). Of course, these designs place specific requirements on substrate and thin film choices.

5.4.5 Software integration

In situ SE can involve communication with the process to provide feedback control. For example, end-point detection can be achieved by sending a signal to close a shutter. Control of temperature or composition may involve feedback to a heater or sputter gun current supply. Software integration can also provide synchronization between the SE and a moving process. Section 5.4.2 already described the benefits of measurement averaging that coincides with beam wobble. Another application of software integration is found in processes that involve movement of the sample within the process. Johs (2004) demonstrated *in situ* monitoring within a PECVD deposition process that rotated both the samples and the planetary table. Due to the non-specular nature of actual parts to be coated, such as gears, a NiCr-plated cylinder was used as witness sample for *in situ* SE. The witness sample rotated around the planetary table at three revolutions per minute, allowing measurement for about 0.5 seconds during each 20 second revolution. High-speed SE measurements at hundreds of wavelengths every 21 milliseconds collected adequate information during this brief encounter with the sample. Measurement triggering was configured to collect data only when adequate detected light intensity reached the SE detector, signifying the witness sample passing through the SE beam. Figure 5.30 shows a schematic of this set-up. Film growth of 5 nm or greater would occur between sample rotations into

130   *In situ* characterization of thin film growth

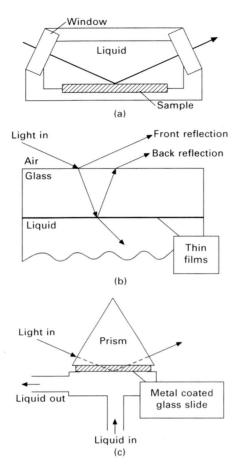

5.29 Three liquid cell geometries to study the interface between a surface and liquid: (a) standard geometry where light enters windows and travels through the liquid to reach the surface, (b) reverse-side ellipsometry through a glass substrate where light does not have to travel through the liquid (adapted from Zudans *et al.*, 2003) and (c) total internal reflection ellipsometry set-up where light enters a prism and does not have to travel through the liquid (adapted from Poksinski and Arwin, 2007).

the SE beam position, but *in situ* SE could still monitor the process and witness optical constant variations in the deposited diamond-like carbon (DLC) films at different stages of film growth.

Software synchronization has also been used to communicate between multiple ellipsometers along the same process chamber used to deposit multilayer coatings. Peros and Kuester (2007) used multiple SE systems to study each layer in multi-layer deposition for solar thermal applications.

In situ spectroscopic ellipsometry (SE)   131

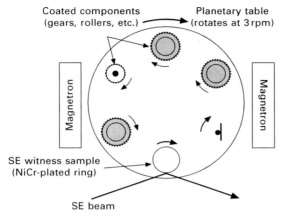

5.30 Top-view of the dual magnetron PECVD sputter chamber with planetary rotation used to coat DLC on gears, rollers, and an SE witness sample. The planetary table rotates at 3 rpm, which positions the witness sample at *in situ* SE measurement location for a short time-period each 20 seconds (adapted from Johs, *Thin Solid Films*, vol. 455–456, pp. 632–638, 2004, with permission).

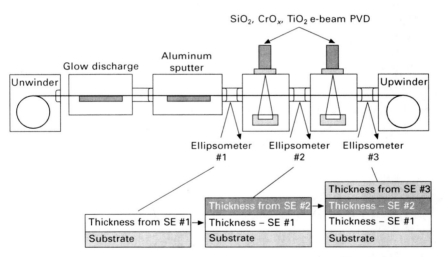

5.31 Roll-to-roll coater incorporating multiple in-line SE systems to determine multi-layer optical coating thicknesses (adapted from Peros and Kuester, 2007, with permission).

Figure 5.31 depicts their roll-to-roll process, where results from the first ellipsometer are passed to the second ellipsometer, and so on. This allows implementation of dynamic process control changes as needed for overall optical coating performance.

132   *In situ* characterization of thin film growth

5.5   Further *in situ* SE examples

In situ applications have been an important field of ellipsometry for at least half a century. The importance of *in situ* SE is reflected by topics of papers given at international conferences on ellipsometry from 1963 to 2010 (Section 5.9.1). At recent conferences, as many as 10% of a few hundred papers involved *in situ* SE. Current *in situ* studies can be loosely grouped into topics such as (i) electrochemistry, (ii) compound semiconductor growth and interfaces, (iii) processes related to uses of silicon in electronics, (iv) biological interfaces, and (v) fundamental surface science. The category we call 'processes related to development of uses of silicon' includes, for example, studies of rapid thermal anneal of ion-implanted crystalline silicon, growth of silicides, control of amorphous to polycrystalline silicon volume composition ratio, or plasma etching dynamics of photoresists on silicon surfaces.

The material systems benefitting from *in situ* diagnostics, and thus topics discussed at conferences, have evolved over the years due to changing commercial interests. For example, *in situ* SE has now been used to study SiGe for circuits; III–V semiconductors for optoelectronics; GaN and related compounds for wide bandgap optical devices; amorphous Si, μc-Si, CdTe or CIGS for photovoltaics; conducting polymers for electronics and photovoltaics; biointerfaces for bio-implant materials and drug testing; and numerous other material systems of commercial importance.

In this section we choose a few specific material and technology areas to describe in detail:

- compound semiconductor growth and process control;
- nanomaterials, nanostructures, and processing for nanotechnology;
- optical coatings and multi-layer structures;
- biomaterials and bio-interfaces;
- *in situ* studies of materials and devices for photovoltaics.

5.5.1   Compound semiconductors

Compound semiconductors exhibit strong absorption above their bandgap energies. Typical wavelengths for band-edge absorption range from the ultraviolet to near-infrared. Optical constant dispersion shapes are affected by temperature, composition, crystallinity, and other material properties. Thus, *in situ* SE can be a useful tool to perform real-time diagnostics of compound semiconductor properties and obtain layer thickness. In fact, in the mid-1990s the promise of real-time monitoring and control for compound semiconductors helped push development of fast, wide spectral range SE systems that could cover appropriate photon energies of semiconductor critical points.

Much of this early work focused on $Hg_{1-x}Cd_xTe$, as maintaining correct detector bandgap and optimum performance for infrared detectors required composition control to better than ±0.002. *In situ* SE has demonstrated this level of performance, as shown in Fig. 5.32 (Phillips *et al.*, 2001) where *in situ* SE composition is compared with *ex situ* Fourier transform infrared (FTIR) spectroscopy measurements for 50 MBE growth runs. The *in situ* SE composition for 90% of the layers in this study fell within ±0.002 of the desired composition, with a 5× improvement in standard deviation for runs with SE control compared with runs without SE control. Composition measured by *in situ* SE depends on well-calibrated optical libraries (Section 5.3.5), which can be produced on the same process chamber later used for wafer production. Phillips *et al.* (2001) went on to demonstrate real-time feedback control of a linear composition profile (Figure 5.33) by adjusting the Te effusion cell temperature based on *in situ* SE composition measurements.

5.5.2 Nanomaterials

Nanomaterials offer novel properties of interest for electronic, optic, photonic, and sensor applications. The high surface sensitivity of SE makes it useful in the study of thin films and structures with nanometer dimensions. Optical

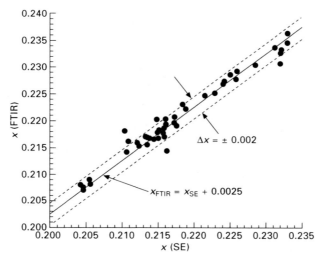

5.32 Results from 50 growth runs show excellent agreement in Cd composition (x) for $Hg_{1-x}Cd_xTe$ thin films determined with *in situ* SE measurements when compared to *ex situ* Fourier transform infrared (FTIR) transmission results. Ninety percent of runs have an accuracy (Δx) within ±0.002 (reprinted with permission from Phillips *et al.*, 2001, *J. Vac. Sci. Technol. B*, vol. 19, no. 4, pp. 1580–1584. Copyright 2001, American Vacuum Society).

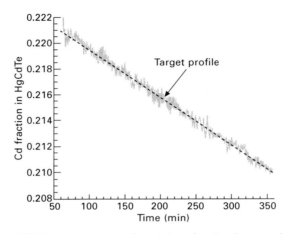

5.33 Demonstration of real-time feedback control to grow a linearly graded HgCdTe film (reprinted with permission from Phillips *et al.*, 2001, *J. Vac. Sci. Technol. B*, vol. 19, no. 4, pp. 1580–1584. Copyright 2001, American Vacuum Society).

properties of these mixed materials have been described using EMA theory (Aspnes, 1982). Here, the volume fraction and shape of each material affect the polarizability of the mixture.

Oates and Christalle (2007) used the Maxwell Garnett EMA to model silver nanoparticles in a poly(vinyl alcohol) (PVOH) host matrix. The optical response from *in situ* SE, specifically a plasmon resonance due to silver nanoparticles, was used to estimate nanoparticle radius. Particle formation for two curing temperatures is compared in Fig. 5.34. The EMA model only considers silver particles as they become large enough to exhibit metallic behavior, which explains why the increasing silver percentage coincides with increasing particle radius.

Some nanostructures, such as nanorods, can exhibit shape anisotropy. EMA theory can describe many of these structures, but must include a directional-dependent parameter to give different polarizabilities parallel and perpendicular to the structures. Hsu *et al.* (2008) demonstrated characterization of silicon nanorods using an anisotropic EMA with variation in void percent as a function of depth into the film. Similar modeling during sputter deposition of GaSb nanopillars led to determination of pillar height and both bottom and top diameters for the pillars (Nerbø *et al.*, 2009). *In situ* SE thickness matched *ex situ* AFM thickness over a range of nanopillar heights from about 20 to 80 nm.

ALD is another candidate for nanotechnology processing; implemented as two self-limiting processes it offers the ability to grow thin films one monolayer at a time. Many researchers are demonstrating the importance of *in situ* SE characterization for ALD processes. Langereis *et al.* (2009)

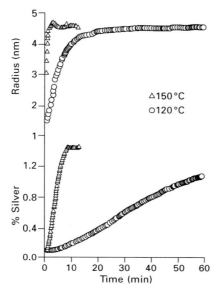

5.34 In situ SE measurements during curing of silver nanoparticles in PVOH host at 120 °C (circles) and 150 °C (triangles) show the increase in silver for the EMA model as the size of nanoparticles increases (adapted from Oates and Christalle, 2007, *J. Phys. Chem. C*, vol. 111, pp. 182–187, with permission from American Chemical Society).

provide an excellent review of *in situ* SE applications for ALD processes and their considerable efforts will not be repeated here; instead we will limit our discussions to a few cases. For example, although ALD is considered an ideal layer-by-layer growth process, nucleation effects can inhibit growth. Figure 5.35 shows ALD growth of a TiN film with poor nucleation on surfaces with low –OH densities, while the TaN shows immediate nucleation (Langereis *et al.*, 2009).

In situ SE monitoring also offers the benefit of tracking variations in growth rates as the film changes. *In situ* SE measurement during ALD of TiO_2 layers show the phase transition between amorphous and crystalline (anatase) film growth, as demonstrated in Fig. 5.36 (Langereis *et al.*, 2009). With these and other similarly important applications, the utility of *in situ* SE continues to expand into new areas of processing for nanostructures.

5.5.3 Optical coatings

Optical coating production often relies on optical monitoring (R/T) or quartz crystal monitoring during thin film deposition. Quartz-crystal monitoring is an inexpensive method of tracking growth rate, but it does not directly measure the sample of interest. R/T sensitivity to very thin layers is greatly

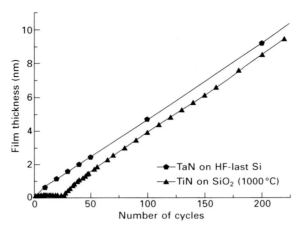

5.35 Film thickness measured by *in situ* SE during ALD growth of TiN and TaN films demonstrating a nucleation effect for the former, but not the latter (reprinted from Langereis *et al.*, 2009, *J. Phys. D: Appl. Phys.*, vol. 42, 073001 (19 pp) with permission from IOP Publishing).

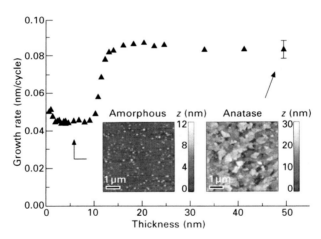

5.36 Growth rate determined from *in situ* SE during plasma-assisted ALD of TiO_2 film showing the shift from amorphous to anatase crystal phase (reprinted from Langereis *et al.*, 2009, *J. Phys. D: Appl. Phys.*, vol. 42, 073001 (19 pp) with permission from IOP Publishing).

diminished, although this is not typically a problem for optical coatings as they often employ quarter-wave optical thickness layers. *In situ* SE can supplement these methods to accurately determine refractive index, detect index gradients, and monitor advanced multi-layer coating structures.

In situ SE measurements are often sensitive to refractive index values to ±0.001, which allows calibration of optical coating processes under various conditions. For example, Larouche *et al.* (2001) studied the index of refraction for TiO$_2$/SiO$_2$ mixtures with varying composition, deposited by PECVD. Figure 5.37 shows the non-linear trend of increasing refractive index with increasing TiO$_2$ composition. Accurate calibration of this deposition behavior allowed the authors to design and fabricate optical coating structures with graded-index.

An example of a graded index optical filter was demonstrated by Amassian *et al.* (2002), as shown in Fig. 5.15. The authors also utilized the sensitivity of *in situ* SE to detect index gradients so they could (i) detect inhomogeneities, (ii) study their origin, and (iii) find process conditions to mitigate their effect on optical design.

In situ SE has also been applied to multilayer stacks. Vergöhl *et al.* (1999) demonstrated control of a four layer anti-reflective coating stack using thin film calculations for all layers in the optical model. As mentioned earlier, this approach is often impractical for graded or multilayer stacks where a large number of layers need to be calculated. Nonetheless, Dligatch *et al.* (2004) applied multi-wavelength *in situ* ellipsometry and photometric monitoring to a multilayer optical coating with 79 layers. They used real-time monitoring to correct for any deviation from the optical design as the layers were being deposited.

5.5.4 Biological films

Use of spectroscopic ellipsometers for biological films has been well-reviewed by Arwin (1998, 2000, 2005, 2011), and is an important topic for

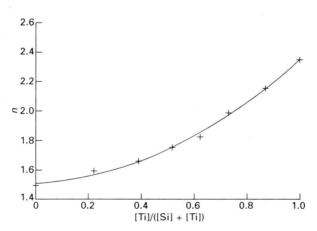

5.37 Calibration of refractive index for PECVD deposited thin film mixtures of TiO$_2$ and SiO$_2$ (reprinted from Larouche *et al.*, 2001, *44th Annual Technical Conference Proceedings-SVC*, pp. 277–281, with permission from the Society of Vacuum Coaters).

overview in this chapter. Most experiments are done in liquids, usually with water as the solvent for proteins and other biomolecules. Common proteins are biotin, avidin, γ-globulin, ferritin, fibronectin, bovine (or human) serum albumin (BSA or HSA), and hemoglobin. Other biomaterials (molecularly more complex) include antigen–antibody bound multilayers, enzymatic catalysists, and enzymes (Berlind et al., 2010).

Most studies have been conducted on smooth interfaces with substrates of crystalline silicon or various metals. Studies have also involved porous silicon surfaces, where the pores offer an increased surface area for reaction, with protein adsorption at the surface as well as penetration of protein into the pores (Karlsson et al., 2005). Berlind et al. (2011) are also testing new surfaces, such as carbon and carbon nitride, in their quest 'for improved materials for life science applications like biomaterials and biosensors'. Unfortunately, the surfaces that are ideal for SE characterization are not often used in practical application such as hip joint implants, heart valves, and stents.

Metal surfaces present a few additional challenges as they are often rough and/or oxidized, and their optical properties vary with both microstructure and surface condition. A common method of addressing these issues is by measuring the substrate prior to bio-film interaction (preferably in liquid solution) to determine reference optical constants for the current surface. Gold, for instance, has no oxide which simplifies analysis, but roughness and microstructure still affect its optical constants. An interface layer can be induced in the near-surface region of gold in response to interaction with an adsorbed biological film, which is speculated to be a region depleted of free electrons (Mårtensson and Arwin, 1995; Mårtensson et al., 1995). This region is optically modeled using an EMA layer to mix the gold and bio-film optical properties.

Gold easily reacts with thiols to allow systematic, fundamental scientific studies of nominally well-oriented proteins with a wide range of chemical species of self-assembled monolayers. Much research to date has been dedicated to fundamental studies of substrate surface chemistry effects, solution composition and concentration, solution pH, temperature, surface hydrophobicity, diffusion of biomolecules from solution, competition between multiple biomolecules simultaneously in solution, and other controllable experimental conditions (Arwin, 2005; Berlind et al., 2008). Most of these studies use 'ideal biomolecules' such as simple or common proteins.

SE measurements are perfectly suited to thin bio-films, as the phase (Δ) is very sensitive to sub-nanometer surface layers. Figure 5.38 shows an in situ SE measurement through a liquid cell during cetyltrimethylammonium bromide (CTAB) adsorption on a gold surface (Tiwald, 2007). CTAB was added to the liquid solution twice during the measurement, followed each time by multiple rinse steps. Kinetic 'Δ' data are shown for a single wavelength of

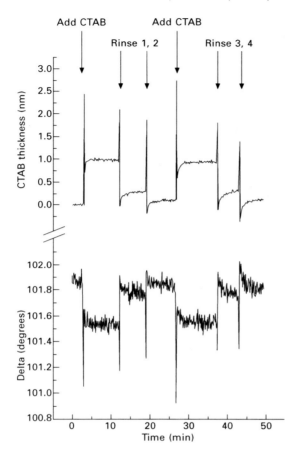

5.38 Measured Δ at wavelength of 600 nm and SE-determined thickness during an *in situ* liquid-cell experiment. CTAB is added to the liquid solution twice, followed by multiple rinse steps (adapted with permission from Tiwald, 2007).

600 nm, while the measurement consisted of 435 simultaneous wavelengths across the visible and near-infrared. Data analysis of the complete spectrum results in a smooth thickness curve (Fig. 5.38), as a result of averaging data from hundreds of simultaneous wavelengths for each time-point.

Because many bio-films are only a few nanometers thick, there is strong correlation between thickness and refractive index which limits the ability to determine both values independently. SE measurements are still very sensitive to the thickness-index product (equivalent to surface mass per unit area), so measurements are often considered without separating these two values. It is most common to fix the index of refraction at measured values from similar thick films. The fixed index allows calculation of an approximate thickness, as with the CTAB in Fig. 5.38 where index was fixed at 1.5. Arwin and Aspnes (1984) successfully separated index and thickness from very thin layers using the substrate spectral response to indicate the correct thickness. However, many bio-films do not meet the 'ideal' requirements for this method to work. Biolayer films at interfaces can (i) lack lateral uniformity because fractional coverage is common over surfaces and complicates data analysis, (ii) be inherently anisotropic, and (iii) configure themselves in various geometries depending on interface surface chemistry, solution pH, surface hydrophobicity, and other factors.

Several promising approaches for bio-film interface studies using *in situ* SE have recently evolved. Measurements at new wavelengths are exploring bio-films from the ultraviolet to the infrared range, where resonant absorptions reveal chemical bonding information (Arwin *et al*., 2008). Total internal reflection ellipsometry (TIRE), also known as surface plasmon resonance enhanced ellipsometry, uses a cell configuration such as shown in Fig. 5.29(c) to provide increased sensitivity to very thin films (Poksinski and Arwin, 2007). Unlike conventional surface plasmon resonance (SPR), which measures reflected intensity, SE uses the full polarization measurement over a wide range of wavelengths and angles to increase sensitivity for a broad range of experimental and surface conditions. Poksinski and Arwin (2004, 2007) have demonstrated very sensitive measurements of protein adsorption using TIRE with thickness resolution as low as 1 picometer, which corresponds to a surface mass density of 100 pg/cm^2. Nabok *et al*. (2005) compared TIRE and conventional SPR for the detection of environmental toxins and found significantly better sensitivity for TIRE measurements.

In the past few years scientists have combined *in situ* SE with the use of quartz crystal monitors (QCM), a long-established method for bio-film diagnostics. QCM technology has been commercialized and widely used for decades. In combinations, *in situ* SE and QCM provide a host of synergistic advantages, since both techniques can simultaneously sense the same film at the same time. This allows researchers to compare results without having to worry about repeatability of film properties sample-to-sample. In addition, SE and QCM report a different thickness (surface mass density); SE reports the total density of attached molecules while QCM reports the total mass coupled to the surface, including both adsorbed molecules and any trapped water. Rodenhausen and Schubert (2011) have proposed methods to interpret the SE and QCM data simultaneously to better understand the formation of adsorbed films. Consider the simultaneous *in situ* SE and QCM measurements during adsorption and rinse for a CTAB film on gold shown in Fig. 5.39 (Rodenhausen *et al*., 2011). The adsorbed film thickness quickly rises upon injection of CTAB with a peak near 14 minutes. Note that, the SE and QCM thicknesses are not equal due to entrapped water, as discussed above. The authors also calculate an absorbent mass fraction, f_o, to relate the QCM and

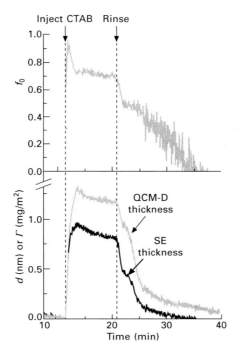

5.39 SE and QCM thickness measured during CTAB adsorption and rinse process. An absorbent mass fraction, f_0 is calculated from both thicknesses to help interpret the film phase evolution (adapted from Rodenhausen *et al.*, 2011, with permission from Elsevier).

SE thicknesses – which is useful when watching the phases of film evolution. During the rinse step, there is a pause suggesting a separation between the removal of weakly bound and strongly bound CTAB molecules at the surface (Rodenhausen *et al.*, 2011).

Another promising area of bio-film interface studies with SE is development of biosensors for rapid testing of new drugs (Arwin, 2005). In sensing applications, surfaces are systemically functionalized with specific chemistries. On-chip optical sensing enables rapid medical analysis using bio-chip arrays. Imaging ellipsometers may also play a role in bio-sensing, but the technique is still in its infancy, especially for practical applications. Lateral resolution has been demonstrated to better than 5 nm (Jin *et al.*, 1996), but most work to date involves single-wavelength ellipsometry. Recent development of imaging SE systems (Jin *et al.*, 2011; Meng and Jin, 2011) may advance the capabilities of such devices, but sampling rates, especially for spectroscopic measurements, are still a limiting factor.

5.5.5 Photovoltaics

In situ SE has long been used as a diagnostic tool for thin film photovoltaic (PV) growth. Much of this work has concentrated on thin film silicon, but recent studies have also involved CdTe, CdS, CdTe$_{1-x}$S$_x$, and CIGS (Li *et al.*, 2007; Sestak *et al.*, 2009). For silicon thin films, *in situ* SE offers the ability to monitor the microstructural phase, surface conditions, and growth rate under various process conditions.

Fujiwara (2007, pp. 311–344) presents an in-depth review of many of these applications, specifically work on the evolution of film growth and surface or interface conditions. Another area of interest for PV films is the measure of crystallinity, which is related to film optical constants and specifically the absorption shape as shown in Figure 5.40 for germanium. Here, the amorphous and nanocrystalline germanium films were prepared with different substrate temperatures during RF magnetron sputtering (Tsao *et al.*, 2010). They are compared with single-crystal germanium, where the prominent optical constant features for crystalline material become broadened

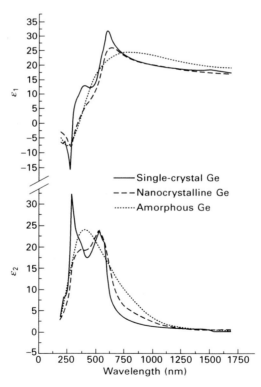

5.40 Optical constants for amorphous and nanocrystalline germanium films compared to crystalline germanium substrate.

with reduced order in the thin films. Thus, SE can monitor the crystalline phase during film growth based on the shape of the film optical constants.

SE has been used to create growth phase diagrams for silicon films deposited by PECVD (Collins et al., 2000; Podraza et al., 2009), hot wire CVD (Levi et al., 2002), VHF PECVD (Cao et al., 2008) and $Si_{1-x}Ge_x$:H films by RF PECVD (Podraza, et al., 2006). Process and sample parameters, such as H_2-dilution gas flow ratio, substrate temperature, and bulk film thickness, are adjusted to determine film properties with *in situ* SE and plot transitions between amorphous, microcrystalline, and mixed-phase growth. Figure 5.41 shows an example of PECVD growth, relating film properties to both H_2-dilution ratio and bulk film thickness, which then allowed the authors to identify the process conditions for optimal PV performance (Podraza et al., 2009). Of specific interest are the film properties of the i-layer within p-i-n or n-i-p silicon solar cells.

5.6 Conclusions

In situ spectroscopic ellipsometry (SE) has been demonstrated to be a versatile diagnostic tool for thin film growth and etch in real-time. Common

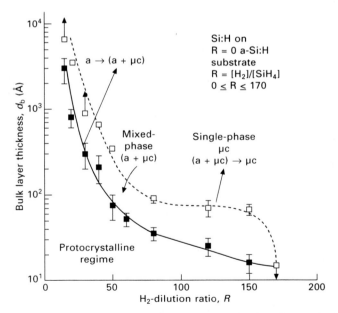

5.41 Phase diagram developed using *in situ* SE for rf PECVD of Si:H thin films with different H_2-dilution ratio and bulk layer thickness (reprinted with permission from Podraza et al., 2009, JVST A, vol. 27, no. 6, pp. 1255–1259, copyright 2009, American Vacuum Society).

measurements include film thickness, surface integrity, growth and etch rate, and optical constants. Because optical constants can be affected by many other material properties, SE measurements can potentially provide valuable information about those as well.

In the past 20 years, *in situ* SE measurements, analysis, and applications have seen significant development and improvement. Spectral range has expanded from laser-based ellipsometry to monochromator-based ellipsometry, with slow spectral resolution, to modern SE systems that can collect over a thousand wavelengths in a fraction of a second. Similarly, accessible wavelength ranges are also expanding, opening new application opportunities.

Other application areas will benefit from the synergistic combination of *in situ* SE with other *in situ* diagnostic methods. This trend can be expected to continue for the foreseeable future, as the combined information is utilized to provide a more complete picture of thin film and deposition properties. As researchers continue to push the boundaries of new applications, they open new doors for industrial integration. This will continue at an increasing rate as *in situ* SE becomes more accessible and the barriers for integration are lowered.

The applications listed in this chapter will soon be supplemented by new areas, as both commercial interests and further improvements expand *in situ* SE into new areas. For example, further development is likely to enable real-time analysis of complex anisotropic materials, nanostructures, and metamaterials by *in situ* SE. This will help improve the ability to monitor such anisotropic films and substrates as chiral nanostructures or multilayer coatings on roll-to-roll plastic processes. After only 20 years, *in situ* SE is gaining maturity and the future holds significant promise for real-time thin film diagnostics.

5.7   Sources of further information and advice

Daraselia, M., Garland, J.W., Johs, B., Nathan, V., and Sivananthan, S. (2001) 'Improvement of the accuracy of the *in-situ* ellipsometric measurements of temperature and alloy composition for MBE grown HgCdTe LWIR/MWIR structures', *J. Electronic Mater.*, vol. 30, no. 6, pp. 637–641.

Duncan, W.M., Bevan, M.J., and Shih, H.D. (1997) 'Real-time diagnostics of II–VI molecular beam epitaxy by spectral ellipsometry', *J. Vac. Sci. Technol. A*, vol. 15, no. 2, pp. 216–222.

Hilfiker, J.N., Glenn, D.W., Heckens, S., Woollam, J.A., and Wierman, K.W. (1996) '*In situ* and *ex situ* optical characterization of electro deposited magneto-optic materials', *J. Appl. Phys.*, vol. 79, no. 8, pp. 6193–6195.

Irene, E.A. (1993) 'Applications of spectroscopic ellipsometry to microelectronics', *Thin Solid Films*, vol. 233, pp. 96–111.

Irene, E.A. (2001) '*In situ* real-time characterization of surfaces and film growth processes via ellipsometry', in *In situ Real-time Characterization of Thin Films*, Auciello, O., and Krauss, A.R. (eds), New York: John Wiley & Sons.

Irene, E.A., and Woollam, J.A. (1995) '*In situ* ellipsometry in microelectronics', *MRS Bulletin*, vol. 20, no. 5, pp. 24–28.

Johs, B., Edwards, J.L., Shiralagi, K.T., Droopad, R., Choi, K.Y., Maracas, G.N., Meyer, D., Cooney, G.T., and Woollam J.A. (1991) 'Real-time analysis of *in-situ* spectroscopic ellipsometric data during MBE growth of III–V semiconductors', *Mat. Res. Soc. Symp. Proc.*, vol. 222, pp. 75–80.

Kildemo, M., Drévillon, B., and Hunderi, O. (1998) 'A direct robust feedback method for growth control of optical coatings by multiwavelength ellipsometry', *Thin Solid Films*, vol. 313–314, pp. 484–489.

Olson, G.L., Roth, J.A., Brewer, P.D., Rajavel, R.D., Jamba, D.M., Jensen, J.E., and Johs, B. (1999) 'Integrated multi-sensor system for real-time monitoring and control of HgCdTe MBE', *J. Electronic Mater.*, vol. 28, no. 6, pp. 749–755.

Roth, J.A., Chow, D.H., Olson, G.L., Brewer, P.D., Williamson, W.S., and Johs, B. (1999) 'Real-time control of the MBE Growth of InGaAs on InP', *J. Crystal Growth*, vol. 201–202, pp. 31–35.

Roth, J.A., Williamson, W.S., Chow, D.H., and Olson, G.L. (2000) 'Closed-loop control of resonant tunneling diode barrier thickness using *in situ* spectroscopic ellipsometry', *J. Vac. Sci. Technol. B*, vol. 18, no. 3, pp. 1439–1442.

Vergöhl, M., Malkomes, N., Matthée, T. and Brauer, G. (2000) 'Real time control of reactive magnetron-sputter deposited optical filters by *in situ* spectroscopic ellipsometry', *Thin Solid Films*, vol. 377–378, pp. 43–47.

Volintiru, I., Creatore, M. and van de Sanden, M.C.M. (2008b) '*In situ* spectroscopic ellipsometry growth studies on the Al-doped ZnO films deposited by remote plasma-enhanced metalorganic chemical vapor deposition', *J. Appl. Phys.*, vol. 103, pp. 033704-1–033704-10.

Woollam, J.A., Johs, B., Herzinger, C.M., Hilfiker, J., Synowick, R., and Bungay, C.L. (1999) 'Overview of variable angle spectroscopic ellipsometry (VASE), Part I: Basic theory and typical applications', *SPIE Proc.*, vol. CR72, pp. 3–28.

Zettler, J.-T. (1997) 'Characterization of epitaxial semiconductor growth by reflectance anisotropy spectroscopy and ellipsometry', *Prog. Crystal Growth and Charact.*, vol. 35, pp. 27–98.

Zudans, I., Heineman, W.R., and Seliskar, C.J. (2004) '*In situ* measurements of chemical sensor film dynamics by spectroscopic ellipsometry. Three case studies', *Thin Solid Films*, vol. 455–456, pp. 710–715.

5.8 Acknowledgments

I wish to thank Professor John A. Woollam for his tireless support and guidance throughout the writing of this chapter. I am also indebted to Professor Hans Arwin for his advice regarding the biological films section. Their contributions are deeply appreciated.

5.9 References

5.9.1 Ellipsometry conference proceedings

Abeles, F. ed. (1983) International Conference for Ellipsometry and other Optical Methods for Surface and Thin Film Analysis, 7–10 June, 1983, Paris, France, *Journal De Physique*, Colloque C10, Supplement no. 12, vol. 44.

Arwin, H., Beck, U. and Schubert, M. eds (2008) 4th International Conference on Spectroscopic Ellipsometry, 11–15 June, 2007, Stockholm, Sweden, *Phys. Stat. Sol. (a)*, vol. 205, no. 4, pp. 709–948, (c), vol. 5, pp. 1003–1442.

Bashara, N.M. and Azzam, R.M.A. eds (1976) Ellipsometry: Proceedings of the Third International Conference on Ellipsometry, 23–25 September, Lincoln, NE, USA, *Surface Science*, vol. 56.

Bashara, N.M., Buckman, A.B. and Hall, A.C. eds (1969) Proceedings of the Symposium on Recent Developments in Ellipsometry, 7–9 August, 1968, Lincoln, NE, USA, *Surface Science*, vol. 16.

Boccara, A.C., Pickering, C. and Rivory, J. eds (1993) Proceedings of the 1st International Conference on Spectroscopic Ellipsometry, 11–14 January, 1993, Paris, France, *Thin Solid Films*, vols. 233–234.

Collins, R.W., Aspnes, D.E. and Irene, E.A. eds (1998) Proceedings of the Second International Conference on Spectroscopic Ellipsometry, 12–15 May, 1997, Charleston, SC, USA, *Thin Solid Films*, vols. 313–314.

Fried, M., Hingerl, K. and Humlíček, J. eds (2004) The 3rd International Conference on Spectrocopic Ellipsometry, 6–11 July, 2003, Vienna, Austria, *Thin Solid Films*, vols. 455–456.

Muller, R.H., Azzam, R.M.A. and Aspnes, D.E. eds (1980) 4th International Conference on Ellipsometry, 20–22 August, 1979, Berkeley, CA, USA, *Surface Science*, vol. 96.

Passaglia, E., Stromberg, R.R. and Kruger, J. eds (1964) *Ellipsometry in the Measurement of Surface and Thin Films*, 5–6 September, 1963, Washington, DC, USA, National Bureau of Standards Miscellaneous Publication 256.

Tompkins, H.G. ed. (2011) 5th International Conference on Spectroscopic Ellipsometry – ICSE-V, 23–28 May 2010, Albany, NY, USA, *Thin Solid Films*, vol. 519, Issue 9.

5.9.2 General references

Adachi, S. (1999), *Optical Constants of Crystalline and Amorphous Semiconductors*, Boston: Kluwer Academic Publishers.

Amassian, A., Larouche, S., Klemberg-Sapieha, J.E., Desjardins, P., and Martinu, L. (2002) 'In situ ellipsometric study of the initial growth stages of a-TiO$_2$ by PECVD', 45th Annual Technical Conference Proceedings-SVC, pp. 250–255.

Arwin, H. (1998) 'Spectroscopic ellipsometry and biology: recent developments and challenges', *Thin Solid Films*, vol. 313–314, pp. 764–774.

Arwin, H. (2000) 'Ellipsometry on thin organic layers of biological interest: characterization and applications', *Thin Solid Films*, vol. 377–378, pp. 48–56.

Arwin, H. (2005) 'Ellipsometry in life sciences', in *Handbook of Ellipsometry*, Tompkins, H.G. and Irene, E.A. (eds) New York: William Andrew.

Arwin, H. (2011) 'Application of ellipsometry techniques to biological materials', *Thin Solid Films*, vol. 519, no. 9, pp. 2589–2592.

Arwin, H, and Aspnes, D.E. (1984) 'Unambiguous determination of thickness and dielectric function of thin films by spectroscopic ellipsometry', *Thin Solid Films*, vol. 113, pp. 101–113.

Arwin, H., Askendahl, A., Tengvall, P., Thompson, D.W., and Woollam, J.A. (2008) 'Infrared ellipsometry studies of thermal stability of protein monolayers and multilayers', *Phys. Stat. Sol. (c)*, vol. 5, no. 5, pp. 1438–1441.

Aspnes, D.E. (1982) 'Optical properties of thin films', *Thin Solid Films*, vol. 89, pp. 249–262.

Aspnes, D.E. (1993) 'Minimal-data approaches for determining outer-layer dielectric responses of films from kinetic reflectometric and ellipsometric measurements', *Appl. Phys. Lett.*, vol. 62, no. 4, pp. 343–345.

Aspnes, D.E. (1996) 'Optical approaches to determine near-surface compositions during epitaxy', *J. Vac. Sci. Technol. A*, vol. 14, no. 3, pp. 960–966.

Aspnes, D.E., Quinn, W.E., Tamargo, M.C., Pudensi, M.A.A., Schwarz, S.A., Brasil, M.J.S.P., Nahory, R.E., and Gregory, S. (1992) 'Growth of $Al_xGa_{1-x}As$ parabolic quantum wells by real-time feedback control of composition', *Appl. Phys. Lett.*, vol. 60, no. 10, pp. 1244–1246.

Azzam, R.M.A., and Bashara, N.M. (1977) *Ellipsometry and Polarized Light*, Amsterdam: Elsevier.

Berlind, T., Pribil, G.K., Thompson, D., Woollam, J.A., and Arwin, H. (2008) 'Effects of ion concentration on refractive indices of fluids measured by the minimum deviation technique', *Phys. Stat. Sol. (c)*, vol. 5, no. 5, pp. 1249–1252.

Berlind, T., Poksinski, M., Tengvall, P., and Arwin, H. (2010) 'Formation and cross-linking of fibrinogen layers monitored with *in situ* spectroscopic ellipsometry', *Colloi and Surfaces B: Biointerfaces*, vol. 75, pp. 410–417.

Berlind, T., Tengvall, P., Hultman, L., and Arwin, H. (2011) 'Protein adsorption on thin films of carbon and carbon nitride monitored with *in situ* ellipsometry', *Acta Biomater*, vol. 7, no. 3, pp. 1369–1378.

Bryne, T.M., Trussler, S., McArthur, M.A., Lohstreter, L.B., Bai, Z., Filiaggi, M.J., and Dahn, J.R. (2009) 'A new simple tubular flow cell for use with variable angle spectrscopic ellipsometry: a high throughput *in situ* protein adsorption study', *Surface Sci.*, vol. 603, pp. 2888–2895.

Cao, X., Stoke, J.A., Li, J., Podraza, N.J., Du, W., Yang, X., Attygalle, D., Liao, X., Collins, R.W., and Deng, X. (2008) 'Fabrication and optimization of single-junction nc-Si:H n-i-p solar cells using Si:H phase diagram concepts developed by real time spectroscopic ellipsometry', *J. Non-Crystalline Solids*, vol. 354, pp. 2397–2402.

Collins, R.W., Koh, J., Ferlauto, A.S., Rovira, P.I., Lee, Y., Koval, R.J., and Wronski, C.R. (2000) 'Real time analysis of amorphous and microcrystalline silicon film growth by multichannel ellipsometry', *Thin Solid Films*, vol. 364, pp. 129–137.

Collins, R.W., An, I., and Chen, C. (2005a) 'Rotating polarizer and analyzer ellipsopmetry', in, *Handbook of Ellipsometry*, Tompkins, H.G. and Irene, E.A. (eds), New York, NY: William Andrew.

Collins, R.W., An, I., Lee, J., and Zapien, J.A. (2005b) 'Multichannel Ellipsometry', in *Handbook of Ellipsometry*, Tompkins, H.G. and Irene, E.A. (eds), New York, NY: William Andrew.

Dligatch, S., Netterfield, R.P., and Martin, B. (2004) 'Application of *in-situ* ellipsometry to the fabrication of multi-layer optical coatings with sub-nanometre accuracy', *Thin Solid Films*, vol. 455–456, pp. 376–379.

Fujiwara, H. (2007) *Spectroscopic Ellipsometry: Principles and Applications*, Chichester: John Wiley & Sons Inc.

Fujiwara, H., Kondo, M., and Matsuda, A. (2003) 'Interface-layer formation in microcrystalline Si:H growth on ZnO substrates studied by real-time spectroscopic ellipsometry and infrared spectroscopy', *J. Appl. Phys.*, vol. 93, no. 5, pp. 2400–2409.

Haberland, K., Hunderi, O., Pristovsek, M., Zettler, J.-T., and Richter, W. (1998) 'Ellipsometric and reflectance-anisotropy measurements on rotating samples', *Thin Solid Films*, vol. 313–314, pp. 620–624.

Hecht, E. (1987), *Optics*, 2nd edition, Reading, MA: Addison-Wesley.

Herzinger, C., Johs, B., Chow, P., Reich, D., Carpenter, G., Croswell, D., and Van Hove, J. (1996) '*In situ* multi-wavelength ellipsometric control of thickness and composition for Bragg reflector structures', in *Diagnostic Techniques for Semiconductor Materials Processing II*, Pang, S.W., Glembocki, O.J., Pollack, F.H., Celii, F.G., and Sotomayor Torres, C.M. (eds), *Mat. Res. Soc. Symp. Proc.*, vol. 406, Pittsburgh, PA pp. 347–352.

Herzinger, C.M., Johs, B., McGahan, W.A., and Woollam, J.A. (1998) 'Ellipsometric determination of optical constants for silicon and thermally grown silicon dioxide via a multi-sample, multi-wavelength, multi-angle investigation', *J. Appl. Phys.*, vol. 83, no. 6, pp. 3323–3336.

Hilfiker, J.N. (1995) Ellipsometric Investigation of the Electrodeposition of Magneto-optic Materials, M.S. Thesis, Electrical Engineering, University of Nebraska.

Hilfiker, J.N. (2005) 'VUV ellipsometry', in *Handbook of Ellipsometry*, Tompkins, H.G., and Irene, E.A. (eds), Norwich, NY: William Andrew, pp. 721–762.

Hilfiker, J.N., Singh, N., Tiwald, T., Convey, D., Smith, S.M., Baker, J.H., and Tompkins, H.G. (2008a) 'Survey of methods to characterize thin absorbing films with spectroscopic ellipsometry', *Thin Solid Films*, vol. 516, pp. 7979–7989.

Hilfiker, J.N., Synowicki, R.A., and Tompkins, H.G. (2008b) 'Spectroscopic ellipsometry methods for thin absorbing coatings', *SVC Proc.*, April, pp. 511–516.

Hsu, S.-H., Liu, E.-S., Chang, Y.-C., Hilfiker, J.N., Kim, Y.D., Kim, T.J., Lin, C.-J., and Lin, G.-R. (2008) 'Characterization of Si nanorods by spectroscopic ellipsometry with efficient theoretical modeling', *Phys. Stat. Sol. (a)*, vol. 205, no. 4, pp. 876–879.

Jellison Jr., G.E. (1999) 'Windows in ellipsometry measurements', *Appl. Optics*, vol. 38, issue 22, pp. 4784–4789.

Jelllison Jr., G.E. (2005) 'Data analysis for spectroscopic ellipsometry', in *Handbook of Ellipsometry*, Tompkins, H.G. and Irene, E.A. eds. New York, NY: William Andrew.

Jellison Jr., G.E., and Modine, F.A. (2005) 'Polarization modulation ellipsometry', in *Handbook of Ellipsometry*, Tompkins, H.G. and Irene, E.A. (eds), Norwich, NY: William Andrew.

Jin, G., Jansson, R., and Arwin, H. (1996) 'Imaging ellipsometry revisited: developments for visualization of thin transparent layers on silicon substrates', *Rev. Sci. Instrum.*, vol. 67, no. 8, pp. 2930–2936.

Jin, G., Meng, Y.H., Liu, L., Niu, Y., Chen, S., Cai, Q., and Jiang, T.J. (2011) 'Development of biosensor based on imaging ellipsometry and biomedical applications', *Thin Solid Films*, vol. 519–519, no. 9, pp. 2750–2757.

Johs, B.D. (1999) System and method for directing electromagnetic beams, US Patent #5,929,995.

Johs, B. (2000) Methods for uncorrelated evaluation of parameters in parameterized mathematical equations for window retardance, in ellipsometer and polarimeter systems, US Patent #6,034,777.

Johs, B. (2004) 'General virtual interface algorithm for *in situ* spectroscopic ellipsometric data analysis', *Thin Solid Films*, vol. 455–456, pp. 632–638.

Johs, B., and He, P. (2011) 'Substrate wobble compensation for *in situ* spectroscopic ellipsometry measurements', *J. Vac. Sci. Technol. B*, vol. 29, no. 3, pp. 03C111-1-5.

Johs, B., Doerr, D., Pittal, S., Bhat, I.B., and Dakshinamurthy, S. (1993) 'Real-time monitoring and control during MOVPE growth of CdTe using multiwavelength ellipsometry', *Thin Solid Films*, vol. 233, pp. 293–296.

Johs, B., Hale, J., and Hilfiker, J. (1997) 'Real-time process control with *in situ* spectroscopic ellipsometry', *III–Vs Review*, vol. 10, no. 5, pp. 40–42.

Johs, B., Hale, J., Herzinger, C., Doctor, D., Elliott, K., Olson, G., Chow, D., Roth, J., Ferguson, I., Pelczynski, M., Kuo, C.H., and Johnson, S. (1998a) 'Real-time monitoring of semiconductor growth by spectroscopic ellipsometry', in *In Situ Process Diagnostics and Intelligent Materials Processing*, Rosenthal, P.A., Duncan, W.M., and Woollam, J.A., (eds), Pittsburg, PA, *Mater. Res. Soc. Symp. Proc.*, vol. 502, pp. 3–14.

Johs, B., Herzinger, D., Dinan, J.H., Cornfeld, A., Benson, J.D., Doctor, D., Olson, G., Ferguson, I., Pelczynski, M., Chow, P., Kuo, C.H., and Johnson, S. (1998b) 'Real-time monitoring and control of epitaxial semiconductor growth in a production environment by *in situ* spectroscopic ellipsometry', *Thin Solid Films*, vol. 313–314, pp. 490–495.

Johs, B., Woollam, J.A., Herzinger, C.M., Hilfiker, J., Synowicki, R., and Bungay, C.L. (1999) 'Overview of variable angle spectroscopic ellipsometry (VASE), Part II: Advanced applications', *SPIE Proc.*, vol. CR72, pp. 29–58.

Johs, B., Hale, J., Ianno, N.J., Herzinger, C.M., Tiwald, T., and Woollam, J.A. (2001) 'Recent developments in spectroscopic ellipsometry for *in situ* applications', *SPIE Proc.*, vol. 4449, pp. 41–57.

Karlsson, L.M., Schubert, M., Ashkenov, N., and Arwin, H. (2005) 'Adsorption of human serum albumin in porous silicon gradients monitored by spatially-resolved spectroscopic ellipsometry', *Phys. Stat. Sol. (c)*, vol. 2, no. 9, pp. 3293–3297.

Kildemo, M., Deniau, S., Bulkin, P., and Drévillon, B. (1996) 'Real time control of the growth of silicon alloy multilayers by multiwavelength ellipsometry', *Thin Solid Films*, vol. 290–291, pp. 46–50.

Kim, S., and Collins, R.W. (1995) 'Optical characterization of continuous compositional gradients in thin films by real time spectroscopic ellipsometry', *Appl. Phys. Lett.*, vol. 67, no. 20, pp. 3010–3012.

Kokkoris, G., Vourdas, N., and Gogolides, E. (2008) 'Plasma etching and roughening of thin polymeric films: a fast, accurate, *in situ* method of surface roughness measurement', *Plasma Processes and Polymers*, vol. 5, pp. 825–833.

Kouznetsov, D., Hofrichter, A., and Drévillon, B. (2002) 'Direct numerical inversion method for kinetic ellipsometric data. I. Presentation of the method and numerical evaluation', *Appl. Opt.* vol 41, no. 22, pp. 4510–4518.

Langereis, E., Heil, S.B.S., Knoops, H.C.M., Keuning, W., van de Sanden, M.C.M., and Kessels, W.M.M. (2009) '*In situ* spectroscopic ellipsometry as a versatile tool for studying atomic layer deposition', *J. Phys. D: Appl. Phys.*, vol. 42, 073001 (19pp).

Larouche, S., Amassian, A., Gujrathi, S.C., Klemberg-Sapieha, J.E., and Martinu, L. (2001) 'Multilayer and inhomogeneous optical filters fabricated by PECVD using titanium dioxide and silicon dioxide', *44th Annual Technical Conference Proceedings-SVC*, pp. 277–281.

Levi, D.H., Nelson, B.P., and Perkins, J.D. (2002) 'Mapping the phase-change parameter space of hot-wire CVD Si:H films using *in-situ* real time spectroscopic ellipsometry', *Mat. Res. Soc. Symp. Proc.*, vol. 715, pp. A25.2.1–6.

Li, J., Stoke, J.A., Podraza, N.J., Sainju, D., Parikh, A., Cao, X., Khatri, H., Barreau, N., Marsillac, S., Deng, X., and Collins, R.W. (2007) 'Analysis and optimization of thin film photovoltaic materials and device fabrication by real time spectroscopic ellipsometry', in *Photovoltaic Cell and Module Technologies*, von Roedern, B. and Delahoy, A.E. (eds), *Proc. of SPIE*, vol. 6651, pp. 665107-1–14.

Liphardt, M.M. (2011) Calculation of beam deviation from sample tilt (personal communication, 19 January 2011).

Martensson, J., and Arwin, H. (1995) 'Interpretation of spectroscopic ellipsometry data on protein layers on gold including substrate–layer interactions', *Langmuir*, vol. 11, pp. 963–968.

Martensson, J., Arwin, H., Nygren, H., and Lundström, I. (1995) 'Adsorption and optical properties of ferritin layers on gold studied with spectroscopic ellipsometry', *J. Colloid Interface Scie.*, vol. 174, pp. 79–85.

Maynard, H.L., Layadi, N., and Lee, J.T.C. (1998) 'Plasma etching of submicron devices: *in situ* monitoring and control by multi-wavelength ellipsometry', *Thin Solid Films*, vol. 313–314, pp. 398–405.

Meng, Y.H., and Jin, G. (2011) 'Rotating compensator sampling for spectroscopic imaging ellipsometry', *Thin Solid Films*, vol. 519, no. 8, pp. 2742–2745.

Nabok, A.V., Tsargorodskaya, A., Hassan, A.K., and Starodub, N.F. (2005) 'Total internal reflection ellipsometry and SPR detection of low molecular weight environmental toxins', *Appl. Surface Sci.*, vol. 246, pp. 381–386.

Nerbø, I.S., Le Roy, S., Kildemo, M., and Søndergård, E. (2009) 'Real-time *in situ* spectroscopic ellipsometry of GaSb nanostructures during sputtering', *Appl. Phys. Lett.*, vol. 94, 213105 (3pp).

Oates, T.W.H., and Christalle, E. (2007) 'Real-time spectroscopic ellipsometry of silver nanoparticle formation in poly(vinyl alcohol) thin films', *J. Phys. Chem. C*, vol. 111, pp. 182–187.

Palik, E.D. ed. (1998a) *Handbook of Optical Constants of Solids*, San Diego, CA: Academic Press.

Palik, E.D. ed. (1998b) *Handbook of Optical Constants of Solids II*, San Diego, CA: Academic Press.

Palik, E.D. ed. (1998c) *Handbook of Optical Constants of Solids III*, San Diego, CA: Academic Press.

Peros, D., and Kuester, H. (2007) 'Status of industrial PVD on metal strips for optical purposes', presented at International Conference on Metallurgical Coatings and Thin Films, San Diego, California.

Phillips, J., Edwall, D., Lee, D., and Arias, J. (2001) 'Growth of HgCdTe for long-wavelength infrared detectors using automated control from spectroscopic ellipsometry measurements', *J. Vac. Sci. Technol. B*, vol. 19, no. 4, pp. 1580–1584.

Podraza, N.J., Wronski, C.R., and Collins, R.W. (2006) 'Deposition phase diagrams for $Si_{1-x}Ge_x$:H from real time spectroscopic ellipsometry', *J. Non-Crystalline Solids*, vol. 352, pp. 1263–1267.

Podraza, N.J., Li, J., Wronski, C.R., Dickey, E.C., and Collins, R.W. (2009) 'Analysis of controlled mixed-phase (amorphous+microcrystalline) silicon thin films by real time spectroscopic ellipsometry', *JVST A*, vol. 27, no. 6, pp. 1255–1259.

Poksinski, M., and Arwin, H. (2004) 'Protein monolayers monitors by internal reflection ellipsometry', *Thin Solid Films*, vol. 455–456, pp. 716–721.

Poksinski, M., and Arwin, H. (2007) 'Total internal reflection ellipsometry: ultrahigh sensitivity for protein adsorption on metal surfaces', *Optics Letters*, vol. 32, no. 10, pp. 1308–1310.

Pribil, G.K., Johs, B., and Ianno, N.J. (2004) 'Dielectric function of thin metal fims by combined *in situ* transmission ellipsometry and intensity measurements', *Thin Solid Films*, vol. 455–456, pp. 443–449.

Rodenhausen, K.B., and Schubert, M. (2011) 'Virtual separation approach to study porous ultra-thin films by combined spectroscopic ellipsometry and quartz crystal microbalance methods', *Thin Solid Films*, vol. 519, no. 9, pp. 2772–2776.

Rodenhausen, K.B. Guericke, M., Sarkar, A., Hofmann, T., Ianno, N., Schubert, M., Tiwald, T.E., Solinsky, M., and Wagner, M. (2011) 'Micelle-assisted bilayer formation of cetyltrimethylammonium bromide thin films studied with combinatorial spectroscopic ellipsometry and quartz crystal microbalance techniques', *Thin Solid Films*, vol. 519, no. 9, pp. 2821–2824.

Sestak, M.N., Li, J., Paudel, N.R., Wieland, K.A., Chen, J., Thornberry, C., Collins, R.W., and Compaan, A.D. (2009) 'Real-time spectroscopic ellipsometry of sputtered CdTe thin films: effect of Ar pressure on structural evolution and photovoltaic peformance', *Mater. Res. Symp. Proc.*, vol. 1165, pp. 393–398.

Snyder, P.G., Woollam, J.A., Alterovitz, S.A., and Johs, B. (1990) 'Modeling $Al_xGa_{1-x}As$ optical constants as functions of composition', *J. Appl. Phys.*, vol. 68, no. 11, pp. 5925–5926.

Tiwald, T.E. (2007) *M-2000/QCM-Z500 measurement of CTAB SAM film on gold* [application note], J.A. Woollam Co.: Lincoln NE, USA.

Tompkins, H.G., and McGahan, W.A. (1999) *Spectroscopic Ellipsometry and Reflectometry: A user's guide*, New York: John Wiley & Sons, Inc.

Tsao, C.-Y. Campbell, P., Song, D., and Green, M.A. (2010) 'In situ low temperature growth of poly-crystalline thin film on glass by RF magnetron sputtering', *Solar Energy Mater. Solar Cells*, vol. 94, pp. 1501–1505.

Urban III, F.K., and Tabet, M.F. (1993) 'Virtual interface method for *in situ* ellipsometry of films grown on unknown substrates', *J. Vac. Sci. Technol. A*, vol. 11, no. 4, pp. 976–980.

Vergöhl, M., Malkomes, N., Staedler, T., Matthée, T., and Richter, U. (1999) '*Ex situ* and *in situ* spectroscopic ellipsometry of MF and DC-sputter TiO_2 and SiO_2 films for process control', *Thin Solid Films*, vol. 351, pp. 42–47.

Volintiru, I., Creatore, M., van Hemmen, J.L., and van de Sanden, M.C.M. (2008a) 'Remote plasma-enhanced metalorganic chemical vapor deposition of aluminum oxide thin films', *Plasma Processes and Polymers*, vol. 5, pp. 645–652.

Woollam, J.A., Gao, X., Heckens, S., and Hilfiker, J.N. (1996) '*In situ* monitor and control using fast spectroscopic ellipsometry', *SPIE Proc.*, vol. 2873, pp. 140–143.

Zudans, I., Seliskar, C.J., and Heineman, W.R. (2003) '*In situ* measurements of sensor film dynamics by spectroscopic ellipsometry. Demonstration of back-side measurements and the etching of indium tin oxide', *Thin Solid Films*, vol. 426, pp. 238–245.

Part III

Alternative *in situ* characterization techniques

6
In situ ion beam surface characterization of thin multicomponent films

L. V. GONCHAROVA, The University of Western Ontario, Canada

Abstract: Two different approaches to investigation of thin film growth using ion beam analysis and spectroscopy are discussed in this chapter. The first is based on the high-resolution variant of classical Rutherford backscattering spectroscopy, and the second combines mass spectroscopy of recoiled ions with ion scattering and direct recoil. The chapter reviews the theoretical background for the techniques and gives relevant experimental set-up details. Several illustrations of the applications are given where these methods are used to study the phenomena occurring during the growth of high-κ dielectric oxides on semiconductor substrates and ferroelectric multicomponent oxide thin film materials.

Key words: electrostatic energy analyzer, time-of-flight (TOF) ion scattering and recoil spectroscopy, initial stages of growth, high-κ dielectrics, ferroelectric oxide thin films.

6.1 Introduction

The pursuit of novel materials and devices strongly depends on existing and new tools for probing their structure, composition and properties at the atomic scale. Although characterization and metrology methods have made great advances, many areas will continue to be challenged by thickness and feature size reduction, and there is still a demand for *in situ*, surface and interface specific analytical tools. These analytical methods (a) should provide a range of surface compositional and structural information on a timescale that is commensurate with the deposition rate, (b) have to be compatible with geometric constraints of the deposition process, and (c) should be functional at the temperatures and pressures required by the thin film growth. Ideally they must also be non-destructive and have high sensitivity. Alternatively, some of these requirements can be bypassed by a fast and reliable way to introduce deposited thin films into the analysis chamber and back, without taking them out of vacuum.

One area where ion beam modification and characterization have played an important role in the past two decades is a new generation of complementary metal oxide semiconductor (CMOS) devices. In particular, the thickness of the gate oxide in CMOS is now in the order of 1 nm. The continued scaling of microelectronic components has made the introduction of new gate materials in CMOS technology necessary (Frank et al., 2009; Schlom et al., 2008). Metal (Hf, Zr, La) oxides, silicates and ternary Hf-based oxides with a dielectric constant higher than that of SiO_2 are being widely investigated as dielectric materials for gate stack and other applications (Demkov and Navrotsky, 2005; Peterson et al., 2004), and devices based on such materials are now commercially available from some manufacturers. These new metal oxide-based materials often have poor electrical performance (i.e., instability of the threshold potential – V_T; Copel, 2008; Schaeffer et al., 2004) believed to be connected to the presence of a large number of Si dangling bonds, traps, and other defects at or near the dielectric/Si interface. The Hf oxide (silicate)/Si interface region is strongly affected by the surface preparation prior to dielectric growth, by growth chemistry, thermal treatment and often also by the nature of the metal gate. There is a large body of work that addresses issues pertaining to interface preparation and characterization using ion beam analysis (Busch et al., 2002; Gustafsson et al., 2006; Goncharova et al., 2006, 2007; Lee et al., 2009, 2010).

Another family of multicomponent oxides, such as $SrBi_2Ta_2O_9$ (SBT), $Ba_xSr_{1-x}TiO_3$ (BST), and $YBa_2Cu_3O_{7-x}$, is remarkable because of their piezoelectric responses (Tenne et al., 2009), ferroelectric polarizability (Zurbuchen et al., 2003; Coondoo et al., 2009), or high temperature superconductivity (Wu et al., 2002). When prepared as thin films, all of them have several different crystallographic phases, with physical properties that are highly dependent on the phase and composition. The challenge of these films is that during vacuum (or oxygen environment) deposition at least one component can be segregated at the surfaces or interfaces. They may also have components that are kinetically stable in a certain concentration range in the film, and this stability depends critically on the substrate temperature and ambient gas composition and pressure.

To address these challenges in multicomponent thin film deposition, several complementary characterization tools are often required. There are several ion beam analysis and spectroscopy methods that can be used to give an exceptionally broad range of surface compositional and structural information either under conditions compatible with the growth environment, or by a simple transfer of thin films in ultra-high vacuum (UHV) from deposition chamber to characterization chamber and back.

This chapter will provide a background discussion of these ion beam analysis (IBA) methods, and give their advantages and limitations. To illustrate the applications of the IBA methods to the study of thin film growth, several examples of thin film materials deposition by atomic layer deposition (ALD) or ion sputter deposition will be presented with the details of *in situ* ion beam characterization.

6.2 Background to ion backscattering spectrometry and time-of-flight (TOF) ion scattering and recoil methods

First, the basic physics of ion beam methods (Alford *et al.*, 2007) needs to be addressed along with the summary of limitations and capabilities specific for each method. The basic scheme of the three methods (ion backscattering spectrometry or, in the low energy range, ion scattering spectroscopy, ISS, direct recoil spectroscopy, DRS, and mass spectroscopy of recoiled ions, MSRI) are illustrated in Fig. 6.1. Medium energy ion scattering (MEIS) is another fairly new and still not very common variant of Rutherford backscattering spectrometry (RBS). MEIS uses the same physical principles as ISS, offers the advantage of superior depth resolution while maintaining the same data interpretation as RBS, and in many cases can give quantitative information about sample composition and surface structure on the nanoscale (Gustafsson, 2009).

6.2.1 Theory of IBA methods

In ion–solid interactions, monoenergetic ions in the incident beam collide with target atoms and are scattered backwards (RBS, MEIS, ISS) or recoil ions forward (DRS, MSRI) into the detector–analysis system, which measures the energies of these particles (Alford *et al.*, 2007; Marchut *et al.*, 1984; Taglauer *et al.*, 1980). In the collision, energy is transferred from the moving particle to the stationary target atom; the reduction in energy of scattered particle depends on the masses of incident and target atoms and provides the signature of the target atoms. The energy transfer or kinematics of inelastic collisions between two isolated particles can be solved fully by applying the conservation of energy and momentum laws.

One can assume that an incident energetic particle of mass M_1 has velocity v_0 and energy E_0, and target atom of mass M_2 is at rest. After the collision,

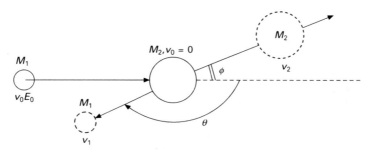

6.1 Schematics of a binary collision process between an incident ion and an atom on the surface under analysis.

the values of the velocities are v_1 and v_2, the projectile and target atom trajectories are determined by the scattering angle θ and recoil angle ϕ. The scattering geometry for the laboratory system of coordinates is presented in Fig. 6.1. Conservation of energy and conservation of momentum parallel and perpendicular to the direction of incidence are expressed by the equations:

$$\frac{1}{2}M_1v^2 = \frac{1}{2}M_1v_1^2 + \frac{1}{2}M_2v_2^2 \qquad [6.1]$$

$$M_1v = M_1v_1\cos\theta + M_2v_2\cos\phi \qquad [6.2]$$

$$0 = M_1v_1\sin\theta - M_2v_2\sin\phi \qquad [6.3]$$

By eliminating the scattering angle, ϕ, first and then v_2, the ratio of projectile energy after collision to incident energy can be given by

$$\frac{E_1}{E_0} = \left[\frac{(M_2^2 - M_1^2\sin^2\theta)^{1/2} + M_1\cos\theta}{M_2 + M_1}\right]^2 \qquad [6.4]$$

Note that mass resolution is the best when the energy transfer from the ion to the target is large. This occurs when the scattering angle is large (small impact parameter), or when the masses of the ion and target are similar.

Other fundamentally important concepts to take into consideration are channeling and blocking phenomena. The principle of channeling and blocking effects can be understood from Fig. 6.2. Typically the incident ion beam is aligned with a high symmetry (channeling) direction of the single crystal substrate. Ions backscatter from the surface atoms, and from the few layers just below the surface, because of the disorder in near-surface region and thermal vibrations. Because of the shadowing effects of the surface atoms, the probability of striking the atom away from the surface plane decreases rapidly. The backscattered flux from the outermost layer would be featureless; however the flux from the subsurface layers will show dips (so-called blocking dips) in directions aligned with the locations of the subsurface atoms. If the surface layer is displaced outwards (or inwards), the surface blocking dip will be observed at larger (smaller) scattering angle (van der Veen, 1985). Determination of the blocking dip position can be done with an accuracy of ~0.1°, which ultimately gives accuracy in the determination of the surface layer displacement of 0.1–0.01 Å. In practice, for most of the cases, the ion energy resolution is often not good enough to allow the separation of the signals from the different surface layers, and cumulative signal for all near-surface layers has to be interpreted.

One of the important distinctions of ion scattering from several top-most surface-sensitive techniques is that incident ions can penetrate and backscatter from the atoms that are relatively deep in the bulk of the material. When incident ions (such as p^+ or He^+) move through matter, they lose energy

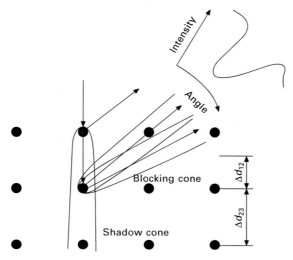

6.2 Schematic illustration of blocking and formation of the blocking dip in the angular spectrum. A change in interlayer spacing is observed experimentally as a shift in the angular position of the blocking dip. The first-to-second and second-to-third interlayer spacings are denoted by d_{12} and d_{23}.

through interactions with electrons that are raised to the excited states, or ejected from atoms. Microscopically, energy losses due to excitations and ionizations are discrete processes. Macroscopically, one can assume that moving ions lose energy continuously. Theoretical treatments of inelastic collision of charged particles with target atoms are separated into 'fast' and 'slow' collision. The criterion is the velocity of the projectile relative to the mean orbital velocity of the atomic electrons in the shell or subshell of a given target atom. Figure 6.3 shows the energy dependence of the stopping power for p^+ and He^+ in Al (Busch, 2000). An estimate of transition velocity between the slow and fast collision cases is the Bohr electron velocity, and this velocity is equivalent of that of 25 keV p^+, or 100 keV He^+. Incident ions with the energies close to the maximum electronic stopping can potentially give better energy resolution. This aspect will be discussed further in Section 6.3.

Another ion beam technique, direct recoil spectroscopy, is less well known than forward ion scattering, but it has been in use in a number of laboratories for several years, especially after recent improvements in time-of-flight (TOF) instrumentation were introduced (Krauss *et al.*, 1995). Direct recoil spectroscopy may be thought as a low-energy equivalent of elastic recoil detection (ERD) (Assmann *et al.*, 1996; Doyle and Peercy, 1979) where grazing incident beam direction and exit angles are used to eject surface atoms in a single collision event with the primary ion. The recoiled surface

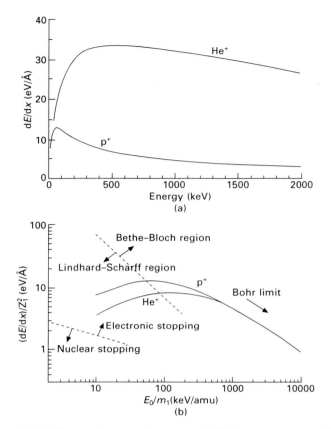

6.3 (a) Electronic stopping power (dE/dx) for protons and He$^+$ ions in Al versus energy. (b) The same data points are plotted on reduced axes which demonstrate the Bohr regime at high energy. The boundary between the Lindhard–Scharff and Bethe–Bloch regimes is shown. The regions in which the total stopping power is dominated by nuclear stopping and electronic stopping are shown.

atoms have sufficiently high energy to be detected by a particle counter such as a channel plate multiplier, and the beam forming and detection hardware is very similar to that of the forward scattering. The main advantage of DRS compared with ISS is the ability of the former method to detect all atomic species, including those that are lighter than the primary one.

DRS is based on detection of direct recoil-sputtered surface atoms which are ejected as the result of a single collision between the primary ion and a surface atom. The kinetic energy E_2 of these atoms is given by

$$\frac{E_2}{E_0} = \frac{4 M_1 M_2}{(M_1 + M_2)^2} \cos^2\phi \qquad [6.5]$$

DRS in fact is one of relatively few surface analytical methods to have good sensitivity for hydrogen and its isotopes (Reichelt et al., 2002).

Mass spectrometry of recoiled ions can be compared with DRS in which only ion fraction of the direct recoil spectrum is detected. It is analogous to a secondary-ion mass spectrometry (SIMS) except that the scattering geometry is optimized for a single collision ejection event, rather than multiple collision cascades associated with SIMS detection. As a result the kinetic energy and ion fraction of the ejected surface atoms are much higher than is the case for SIMS.

6.3 Experimental set-ups

To produce incident ions all of the above methods will rely on the high-voltage ion beam sources, implanters, or even accelerators (Hellborg, 2009). General system design can vary dramatically from the large-scale installations based on several MeV electrostatic accelerators to more compact footprint vacuum chambers, with the ion source in a close proximity to the analyzed target. For the purpose of thin film analysis discussion in this chapter, it is important to note that for *in situ* thin film analyses, especially when atomic layer or nanoscale resolution is critically important, requirements for the detector are stricter. Among high-resolution detectors introduced in the past two decades, two detector types – electrostatic energy analyzer (ESA) (Turkenburg et al., 1976) and TOF – have been used most often in the *in situ* deposition and analysis system, and will be discussed here in detail.

One large electrostatic ion energy analyzer for high-resolution MEIS measurements was developed in FOM (Institute voor Atom- en Molecuulfysica, the Netherlands) (Turkenburg et al., 1976). Data acquisition for multiple detection angles is crucial in many cases, therefore a spectrometer simultaneous detection of ion intensity at many different angles was designed by Tromp et al. (1984). A new version of ESA was constructed in the early 1980s and represents the equipment used by large part of the community today. The design was later modified by removing the exit slit and inserting a two-dimensional detector in its place. As a result, angular information is obtained on one coordinate, and energy on the other. The final toroidal ESA design (Fig. 6.4) was proposed and is now commercially available from High Voltage Engineering in the Netherlands (http://www.highvolteng.com/).

In the toroidal ESA, the width of the entrance slit controls the accepted angular range, and the height of the beam spot is reduced so to give good energy resolution. Ions that scatter from the surface with the same energy and parallel directions, but at different points in the horizontal direction, will be focused on the same point on the position-sensitive detector. To analyze the energy and angle of the ions passing through the toroidal ESA, the detector utilizes a pair of micro-channel plates in a chevron mounting, followed by a

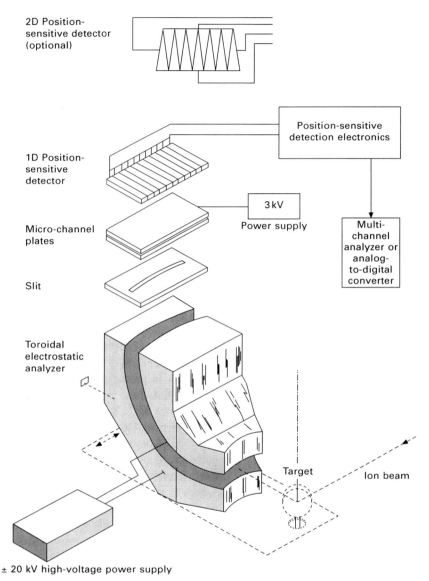

6.4 Toroidal ion energy analyzer (High Voltage Engineering, Amersfoort, The Netherlands). If the optional 2D position-sensitive detector is used, no exit slit is used.

multi-anode charge-dividing collector (Busch, 2000). Other researchers have explored different kinds of ESA with better energy resolution but they are typically less flexible with regards to the scattering geometry. Toroidal ESA offers many advantages; however, its main limitation is the micro-channel plates that are highly sensitive to the high-pressure spikes. The bulky design of the detector and the need to move and to position it at different scattering angles, ϕ (in $-90° < \phi < +180°$ continuous range) on a rotation table inside a scattering chamber, almost completely exclude the possibility of differential pumping.

In order to investigate the ALD growth mechanism of high-κ dielectric thin films during the initial growth stages, the samples can be transferred *in situ* from the deposition chamber with pressures in the $0.1-10^{-5}$ torr range to the adjacent ultra-high vacuum MEIS chamber using vacuum transfer mechanisms, after each cycle of deposition. Several research groups (Chang *et al.*, 2005; Chung *et al.*, 2007; Copel, 2008; Lee *et al.*, 2009) have utilized this approach. After MEIS analyses are completed, sample can be transferred back to the ALD chamber for the following deposition cycle.

Ion beam damage to the sample is another factor that should be considered in MEIS depth profiling on semiconductor substrates. Unlike many metals, which are 'self-annealing' at moderate temperatures, atoms in insulators or semiconductors have lower mobility. Typically double channeling alignment is used in MEIS, where both incident beam and the center of the detector are aligned to the major crystallographic directions of the substrate. If the background yield increases due to the scattering from newly formed defects, quantification of the light elements becomes accurate, since their peaks are always detected on top of the substrate background yield. Fortunately, for ~100 keV protons beam damage is easy to control. For a typical ion dose (~10^{15} ions/cm^2) it is small (~ five displaced atoms in a thousand). If higher ion beam dose is required to finish the measurements, the ion beam has to be moved to a different spot on the sample surface.

Another *in situ* integration approach has been developed by Argonne National Laboratory group (Auciello *et al.*, 1996; Im *et al.*, 1996; Krauss *et al.*, 1995, 1998). The three complementary pulsed ion beam surface analytical techniques ISS, DRS and MSRI collectively represent TOF ion scattering and recoil spectroscopy system (Im *et al.*, 1996; Krauss *et al.*, 1998), shown in Fig. 6.5. Mass spectroscopy of recoiled ions is a form of forward recoil spectroscopy that uses time refocusing analyzers (Poschenrieder and Oetjen, 1972) to eliminate the multiple scattering background of the direct recoil spectrum, obtaining much greater sensitivity and greater mass resolution than the simple line of sight direct recoil detector. The differences between mass spectroscopy of recoiled ions and SIMS are summarized in the review by Krauss *et al.* (1994).

In this set-up (Krauss *et al.*, 1998), the scattering/deposition chamber is equipped with ion sputtering gun focused onto a target mounted on a rotatable carousel with several targets. Primary beam for the ion beam analysis is produced by a telefocused ion source. The highly collimated beam passes through two deflection regions, spaced from each other by a

164   *In situ* characterization of thin film growth

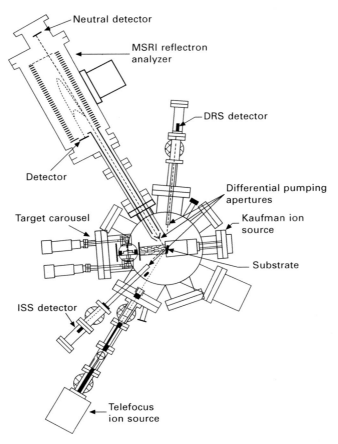

6.5 Schematic of the TOF–ion scattering and recoil spectroscopy system constructed by Krauss *et al.* (1998), indicating the location of the ion beam line, and ISS, DRS, and MSRI detectors.

drift region, and isolated from the main scattering chamber by an aperture. By controlling the time interval between the applications of voltage steps to the two sets of deflection plates, a pulsed ion beam with pulsed duration adjustable in the 10–1000 ns range is produced. A final set of direct current deflection plates is then utilized to steer the beam to the desired location on the sample. All beam lines are differentially pumped by turbomolecular pumps. Two channel electron multiplier detectors are positioned at angles of 165° and 25° relatively to the ion beam direction, and the reflectron for the MSRI analyzer is placed at a scattering angle of 60° (Krauss *et al.*, 1998).

Each detector is differentially pumped and sees the sample through a 1 mm diameter aperture. All detectors and associated optics are located sufficiently far away from the sample so that they do not block the deposition flux. By differentially pumping the space between the apertures, it is possible to maintain a detector vacuum approximately three orders of magnitude lower than that of the sample environment.

6.4 Studies of film growth processes relevant to multicomponent oxides

Ferroelectric perovskite oxide thin films such as Pb(Zr$_x$Ti$_{1-x}$)O$_3$ (PZT) and SrBi$_2$Ta$_2$O$_9$ (SBT) have been studied as potential candidates for non-volatile ferroelectric random access memories (NVFRAM). Because of the high crystallization temperature of ferroelectric SBT films (>700°C), SBT integration with Si-based technology requires a stable metal barrier heterostructure. One possible stack structure for a bottom electrode is Pt/Ti/SiO$_2$/Si thin film. Pt is often used as the capacitor electrode material because of its high electrical conductivity and because Pt lattice parameters give a good epitaxial relationship for the ferroelectric layer growth. The main functions of Ti are to improve adhesion of Pt to SiO$_2$, and prevent Pt direct contact with silicon to form Pt silicides. Yet the previous studies of Pt/Ti layer stacks in an oxygen atmosphere at high temperatures showed their instability. During heating Ti diffused into the Pt layer along the grain boundaries and became incorporated in the SBT layer (Sreenivas et al., 1994). Ti interactions with SBT layer in turn affect composition, microstructure and electrical properties of the SBT-based capacitors.

In case of Pt/Ti/SiO$_2$/Si stack, a relatively poor depth resolution of RBS limits information on whether the diffusion process results in the appearance of Ti atoms on the topmost Pt surface. The latter is not a desirable scenario for the initial stages of SBT film growth. Mass spectroscopy of recoiled ions (Krauss et al., 1998) has higher depth resolution than RBS, and it was therefore used to obtain a detailed picture of the surface Pt/Ti/SiO$_2$/Si composition (Fig. 6.6). One can see that the surface composition in vacuum is dominated by H, C, O, Pt, and Ar, and small amount of Ti. The relative chemical inertness and high mass of Pt results in lower sensitivity of Pt detection than that for Ti, Si, and H. As the temperature increases, the Pt signal decreases and nearly disappears, while at the same time the Ti signal increases significantly and a large Si peak appears. It is estimated that after 700°C anneal, at least 80% of the surface consists of Ti and Si. It is interesting that if a similar annealing procedure (700°C) is conducted in an oxygen ($P_{O_2} = 5 \times 10^{-4}$ torr) environment, only a small Si signal is detected on the surface. It has been proposed by Bruchhaus et al. (1992) that this may be due to the TiO$_x$ diffusion barrier formation at the Ti/Si interface as a result of oxygen diffusion via Pt grain boundaries.

Diffusion processes are also extremely important in novel high-κ dielectric material structures. When different metals are deposited on a HfO$_2$ (or Hf

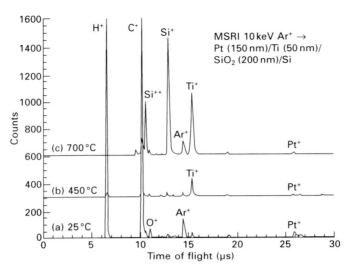

6.6 MSRI spectra obtained by Krauss *et al*. (1998) for a 10 keV Ar⁺ primary ion incident on multilayered structure with Pt (150 nm)/Ti (50 nm)/SiO$_2$ (200 nm)/Si. Note that the signal intensities in the MSRI spectra do not directly correspond to the surface coverage.

silicate)/SiO$_2$/Si stack, additional interactions may occur both at the metal/high-κ and high-κ/Si interfaces, which may either improve or degrade electrical device performance. Several metals have been investigated with the goal of improving electrical behavior of the overall gate stack. It was shown by Preisler *et al.* (2004) that in W/HfO$_2$ gate stacks the tungsten can under certain conditions introduce excess oxygen, resulting in interfacial SiO$_2$ growth at the HfO$_2$/Si interface. It was suggested that Ti acts as an oxygen getter when used as a gate metal on HfO$_2$/SiO$_2$ and ZrO$_2$/SiO$_2$ dielectric stacks (Kim *et al.*, 2004). Using *in situ* Ti deposition and MEIS measurements, it was shown that Ti metallization of HfO$_2$/SiO$_2$/Si stacks reduces the SiO$_2$ interlayer and (to a more limited extent) the HfO$_2$ layer. MEIS backscattering spectra for an as-deposited HfO$_2$/SiO$_2$/Si stack after 4.2 nm Ti deposition *in situ* and after a subsequent anneal to 300 °C in UHV are presented in Fig. 6.7 (Goncharova *et al.*, 2007). After Ti deposition, the Hf, Si and interface O peaks are shifted from their surface energy peak positions to lower energies, due to energy loss in the Ti layer. Simulations for an HfO$_2$/SiO$_2$/Si stack after Ti deposition (Fig. 6.7b) confirm that the Ti layer thickness is uniform. A distinct surface O peak (~106 keV) and the low O yield between the surface and interface peaks (106–107 keV) indicate that the oxygen concentration in the Ti layer is small (Ti-O$_x$, $x < 0.10$), with enhanced oxidation occurring at the surface (2–3 Å of Ti$_2$O$_3$) and at the Ti/HfO$_2$ interface (5 Å TiO$_x$, $0.5 < x < 1$). From MEIS spectra simulations, it was also concluded that Si atoms

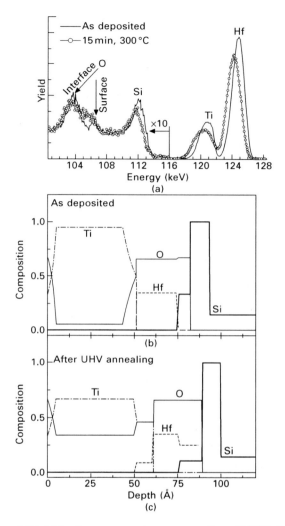

6.7 (a) Medium energy ion spectra for a Ti/HfO$_2$/SiO$_2$/Si(001) stack directly after metal deposition (solid line) and subsequent to a 300 °C UHV anneal (open symbols). (b) and (c) Best fit depth profiles for Ti, Hf, Si and O (adapted from Goncharova *et al.*, 2007).

initially present in the interfacial SiO$_2$ layer incorporate into the bottom of the high-κ layer. Some evidence for Ti–Si interdiffusion through the high-κ film in the presence of a Ti gate in the crystalline HfO$_2$ films has also been reported, as illustrated in Fig. 6.8. This diffusion is likely to be related to defects in crystalline HfO$_2$ films, such as grain boundaries.

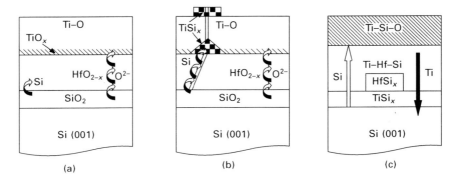

6.8 Schematic showing the proposed O, Si and Ti diffusion processes across (a) as-deposited, (b) crystallized HfO$_2$, following the interfacial SiO$_2$ reduction after 600 °C annealing, (c) after 1000 °C annealing.

6.4.1 *In situ* analysis of Hf oxide growth

The initial stages of HfO$_2$ growth are critical for the development of alternative to SiO$_2$ high-κ materials because dielectric layer properties depend on the control of initial layer growth and the interface composition and defects. The final thickness of Hf oxide or silicate layer can be as small as 20–40 Å; therefore conventional RBS analysis with a resolution of ~100 Å, although still useful to detect total number of Hf atoms deposited at each ALD deposition cycle (Delabie *et al.*, 2005; Wang *et al.*, 2007), is little use in obtaining a quantitative picture of the initial deposition steps, and interface composition. To assist the understanding of each individual parameter (temperature, substrate termination, cycle duration) during ALD growth, MEIS was utilized to measure elemental depth profiles of deposited species after each cycle *in situ* (Chang *et al.*, 2005; Chung *et al.*, 2007). As can be seen from a proton backscattering energy spectra, in Figs 6.9 and 6.10, a gradual increase of Hf peak intensity is well correlated with the Si peak shift to the lower energy. This behavior is consistent with systematic increase of the HfO$_2$ layer thickness. However, ALD growth deviates from ideal linear growth behavior at the initial stage due to the island growth. This can be easily concluded from the comparison of calculated ion yield for the HfO$_2$ ideal monolayer and experimental curve for the same Hf peak areal density (Fig. 6.9). For a more detailed interpretation of the HfO$_2$ growth mechanism, the Hf peak in MEIS energy spectrum can be calculated using simulation software. It was shown by Chang *et al.* that initial interaction between incoming precursors of Hf and oxidizing molecules (H$_2$O, or O$_3$) are strongly dependent on the surface preparation prior to ALD growth: the growth on the H-terminated surface is nonlinear and the growth on the chemical oxide surface is linear (Fig. 6.11). Based on the MEIS results mentioned above, an initial growth model has been proposed (Fig. 6.12).

In situ ion beam surface characterization   169

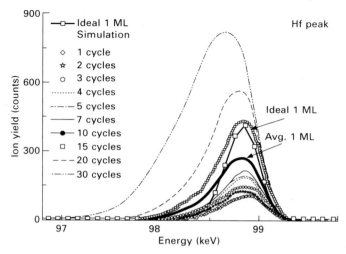

6.9 Medium energy ion scattering spectra of Hf atoms for layers grown on a thermal oxide obtained by the group of Moon (Chang *et al.*, 2005). Atomic layer deposition growth deviates from ideal linear growth behavior at the initial stage because of island-like growth contribution.

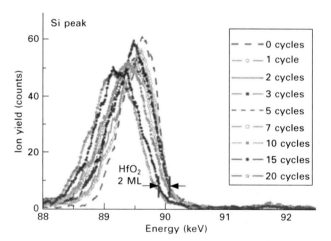

6.10 Medium energy ion scattering spectra of Si atoms for the HfO_2 layers grown on a thermal oxide. After 15 cycles, the leading edge of the Si peak is moved and the first 1 ML is completely formed. A small shoulder indicated by the arrows is observed, which is an indication of slight silicate formation (Chang *et al.*, 2005).

According to this model island-like growth is predominant up to 15 cycles, after which, the first 1 monolayer (ML) forms completely. As a result, island-like growth occurs during the initial HfO_2 growth and islands are merged

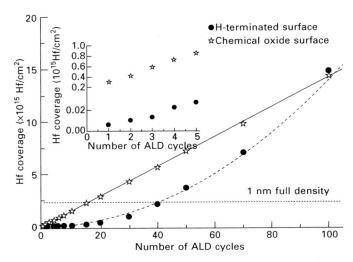

6.11 Hf coverage, measured by *in situ* medium energy ion scattering (Chung *et al.*, 2007), as a function of the number of HfO$_2$ deposition cycles, for the thermal oxide. The inset shows the calculated maximum island height obtained from simulation.

into a continuous atomic layer with increasing ALD cycles, summarized in Fig. 6.12. This study (Chang *et al.*, 2005) was helpful in understanding interfacial reactions that occur during the initial growth stage and to control the reactions for the application of high-κ dielectrics.

6.4.2 Studies of the initial stages of Al oxide film growth on GaAs

Silicon has been the material of choice in the development of CMOS devices for many years, and if it is replaced by other high electron- and hole-mobility semiconductors, such as GaAs or InGaAs, new high-κ dielectric layer and interface passivation composition will have to be established. The quality of Ga and As native oxides is poor for these applications. As a result, it is well known that difficulty in obtaining high performance is related to the challenge of depositing stable passivating layers on GaAs and InGaAs, which leads to high density of defects and Fermi-level pining. All these factors have slowed their integration into mainstream applications. Chemical cleaning and subsequent passivation were used in the past (Kim *et al.*, 2006) to remove the poor-quality GaAs oxide layer. One potentially promising method – the so-called 'self-cleaning' approach – is to choose deposition precursors and conditions so that native Ga and As oxide layer can be removed during initial cycles of dielectric growth. The feasibility of such an approach was illustrated in the case of Al oxide deposition on GaAs surface using ALD

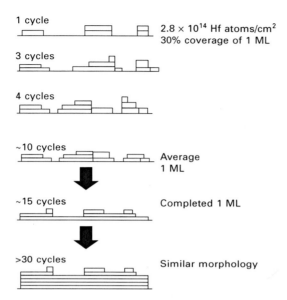

6.12 A model suggested by Chang *et al.* (2005) for the growth mechanism of an ALD HfO$_2$ layer grown on a thermal oxide at the initial growth stage. Island-like growth occurs at the initial stage and the growth then follows a slightly deviated ideal linear growth behavior.

with the trimethylaluminum (TMA) precursor and *in situ* characterization of the growing layer by MEIS and X-ray photoemission spectroscopy (Lee *et al.*, 2009, 2010).

Owing to the similarity in masses, the Ga and As signals overlap in MEIS (Fig. 6.13), yet after TMA exposure, a strong Al signal is observed and the cumulative Ga + As peak is shifted towards lower energy, which will be consistent with the formation of AlO$_x$ film. At the same time there is little change in the oxygen areal density. By analyzing the line shape of all O, Al, Ga and As peaks in MEIS spectrum, Lee *et al.* (2009, 2010) were able to show that initial samples can be modeled as 10 Å thick Ga-rich native oxide layer on top of GaAs substrate, while after 5 Å of Al oxide deposition this interfacial oxide layer is reduced to 7 Å.

Furthermore, stoichiometry of the growing AlO$_x$ layer and interfacial GaAs oxide layer changes as the deposition sequence progresses. Figure 6.14 shows the areal density of Al and O of samples after preheating (320 °C) and after various full cycles of TMA and water. Although the O density changes little initially, the Al density increases a lot (0 to 1.5×10^{15} atom/cm^2), indicating that the oxygen in the Al oxide layer is not from oxygen (water) exposure but from the GaAs native oxide.

172   *In situ* characterization of thin film growth

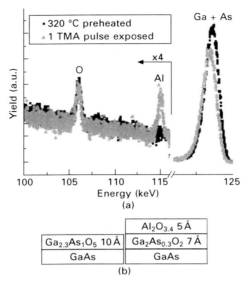

6.13 (a) Medium energy ion scattering spectra for a GaAs sample taken after preheating at 320 °C for ~30 min., and after exposure to 1 trimethylaluminum pulse, and (b) layer models depicting the approximate composition of the films (Lee *et al.*, 2010).

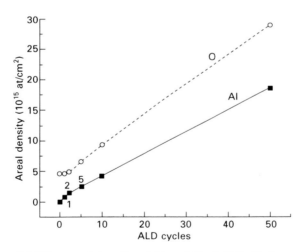

6.14 The areal density of Al and O of Al_2O_3/GaAs samples: after preheating (320 °C), and after 1, 2, 5, 10, and 50 cycles of TMA and water (Lee *et al.*, 2010).

After two cycles a different pattern is observed: both Al and O signals increase linearly with a ratio of 2:3, as expected for a stoichiometric Al_2O_3 film.

6.4.3 Initial stages of ferroelectric perovskite oxide thin film growth

Since it has been reported that Ti degrades the electrical properties of SBT-based capacitors (Schindler *et al.*, 1997), the experiments were focused on studying the initial growth of SBT films on modified Pt surfaces with diffused Ti and Si species, and on alternative stable electrode structures such as $Pt/TiO_2/SiO_2/Si$, $Pt/Ta/TiO_2/SiO_2/Si$, and $RuO_2/Si_2/Si$. These experiments were carried out at 700 °C under $p_{O_2} = 5 \times 10^{-4}$ torr, since these are suitable parameters to deposit SBT films using sputter deposition.

Figure 6.15 shows that when SBT films are deposited at different temperatures in an oxygen atmosphere at $p_{O_2} = 5 \times 10^{-4}$ torr on Pt/Ti/SiO_2/Si multilayered structure, neither Ti nor Si segregation is observed, and there is close correlation between deposition conditions, Bi concentration in the film and electrical properties. If one considers BST crystallographic structure (Krauss *et al.*, 1998), it is possible to see that Bi is either missing from all lattice sites or only selected lattice sites. The MSRI results indicate almost complete loss of Bi during SBT deposition on Ti-containing surface at elevated temperature. Independent ISS study with incident 10 keV Ar^+ primary beam was performed during heating in oxygen up to 570 °C of 44 Å thick SBT film deposited directly on a Ti layer on a MgO substrate. It showed that some of the Bi atoms may be still retained in the film. Ta and

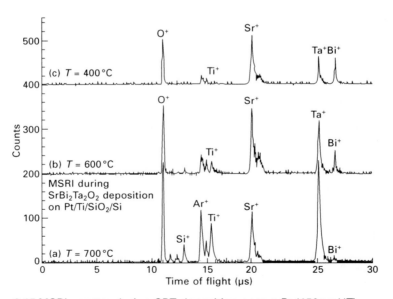

6.15 MSRI spectra during SBT deposition onto a Pt (150 nm)/Ti (50 nm)/SiO_2 (200 nm)/Si substrate at various temperatures in an oxygen background at 5×10^{-4} torr (Krauss *et al.*, 1998).

Sr were clearly seen at room temperature (Fig. 6.16), although it is apparent that the resolution and sensitivity of ISS are not as good as that of MSRI. As temperature increases, first the Sr peak and then the Ta peak disappear, leaving a nearly constant Bi peak and an increasing broad background signal characteristic of multiple scattering from a low mass atom such as O. The removal of the Ta and Sr peaks can be interpreted as due to the movement of Ta and Sr out of the surface layer into the positions in a layered perovskite structure.

The contradiction between the ISS and MSRI results is partially resolved if one considers angular resolved ISS data in Fig. 6.17. As the angle of incidence increases from normal to 22° off normal, the Bi signal disappears and the O multiple scattering peak splits in two. These angular resolved ISS

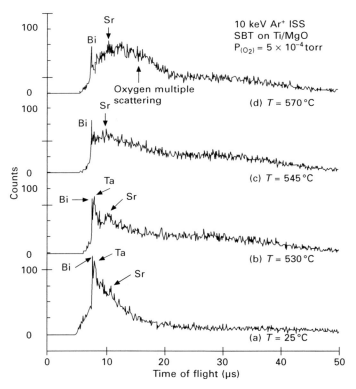

6.16 In situ, real-time ISS spectra of: (a) as-deposited Sr, Bi, Ta species on a room temperature Ti/MgO substrate and during heating at 530 °C (b), 545 °C (c), and 570 °C (d). The evolution of the Sr, Bi, and Ta peaks correlates with the expected formation of the SBT structure, where the Bi atoms remain close to the surface in an incomplete $(Bi_2O_2)_2^+$ layer, while the Sr and Ta atoms move to the perovskite layer underneath the surface (Auciello *et al.*, 1996).

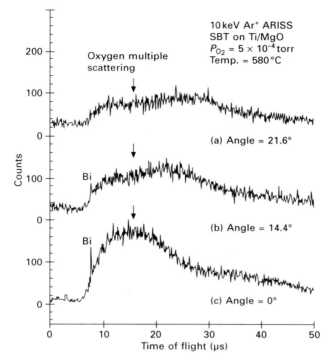

6.17 Angular resolved ISS spectra of the SBT film after the annealing and ISS analysis. The angle indicated in the figure is the angle of incidence of the ion beam with respect to the SBT film surface normal (Auciello *et al.*, 1996).

results are consistent with the film composition where Bi atoms are missing in only the uppermost surface layer, but still present in the deeper layers (below oxygen plane within the crystal structure). These deeper layers are only visible to primary ions that are close to the surface normal. This Bi–O non-stoichiometric ratio at the SBT surface can be related to this material's poor fatigue resistance. The deposited structure is different at the SBT/Pt interface, as confirmed by DRS and MSRI. The DRS oxygen peak intensity obtained during SBT film growth quickly achieves the values corresponding to equilibrium stoichiometric oxygen concentration. This slightly oxygen-rich stoichiometric ratio at the SBT/Pt interface can be correlated with resistance and related to the formation of oxygen vacancies, and better device performances.

Finally, despite the fact that most of the systems mentioned in this chapter were multicomponent oxide thin films and heterostructures, the described ion scattering techniques have been successfully applied to study growth of other materials, such as high-temperature superconductors (Auciello *et al.*, 1995) and diamond-like films (Krauss *et al.*, 1995).

6.5 Conclusions

This chapter demonstrates that several ion beam analysis techniques (MEIS, ISS, DRS and MSRI) are powerful analytical tools for *in situ* characterization of various metal oxide film growth processes. Using MEIS, mechanisms of ALD thin films growth have been revealed for HfO_2/Si and Al_2O_3/GaAs structures. In another example it was shown that the interfacial SiO_2 layer thickness in a HfO_2/SiO_2/Si gate stack is reduced, in some cases with low temperature annealing, following the deposition of a Ti overlayer, known to have a high solubility for oxygen. The HfO_2 layer itself is also reduced in this process. Thermal stability of SBT/Pt/Ti/SiO_2/Si heterostructures has been studied by ISS, DRS and MSRI. These studies demonstrated the ranges of stability of the prepared structures with regard to Ti, Si, Ta segregation and Bi atoms loss.

Some of these segregation or loss phenomena, such as the non-stoichiometry which occurs at the surface of SBT, may be beneficial, although many other phenomena, such as oxygen deficiency at the electrode–ferroelectric interfaces, interlayer diffusion and oxygen gettering-induced destabilization of the more volatile components of the multicomponent oxide ferroelectric materials are detrimental. TOF ion scattering and recoil spectroscopy system analysis during growth represent a unique means of understanding and controlling thin film growth phenomena of complex materials.

6.6 Acknowledgments

I gratefully acknowledge funding provided from NSERC/CRSNG and the University of Western Ontario. I would also like to thank Dr W. N. Lennard for his many insightful suggestions.

6.7 References

Alford T. L., Feldman L. C. & Mayer J. W. (2007) *Fundamentals of Nanoscale Film Analysis*, New York, Springer.

Assmann W., Davies J. A., Dollinger G., Forster J. S., Huber H., Reichelt T. & Siegele R. (1996) ERDA with very heavy ion beams. *Nucl. Instr. Meth. B*, **118**, 242–250.

Auciello O., Krauss A. R. & Schultz J. A. (1995) Time-of-flight ion beam surface analysis for *in-situ* characterization of thin-film growth process. *MRS Bull.*, **20**, 18–23.

Auciello O., Krauss A. R., Im J., Gruen D. M., Irene E. A., Chang R. P. H. & McGuire G. E. (1996) Studies of film growth processes and surface structural characterization of ferroelectric memory-compatible $SrBi_2Ta_2O_9$ layered perovskites via *in situ*, real-time ion-beam analysis. *Appl. Phys. Lett.*, **69**, 2671–2673.

Bruchhaus R., Pitzer D., Eibl O., Scheithauer U. & Hoesler W. (1992) Ion beam analyses of thermal stability Pt/Ti/SiO_2/Si multi-component films. *Mater. Res. Soc. Symp. Proc.*, **243**, 123.

Busch B. W. (2000) Metal and alloy surface structure studies using medium energy ion scattering. PhD thesis, New Brunswick, Rutgers University.

Busch B. W., Schulte W. H., Garfunkel E., Gustafsson T., Qi W., Nieh R. & Lee J. (2002) Oxygen exchange and transport in thin zirconia films on Si(100). *Phys. Rev. B*, **62**, R13 290–293.

Chang H. S., Hwang H., Cho M. H. & Moon D. W. (2005) Investigation of the initial stage of growth of HfO_2 films on Si(100) grown by atomic-layer deposition using *in situ* medium energy ion scattering. *Appl. Phys. Lett.*, **86**, 031906.

Chung K. B., Whang C. N., Chang H. S., Moon D. W. & Cho M. H. (2007) Initial nucleation and growth of atomic layer deposited HfO_2 gate dielectric layers on Si surfaces with the various surface conditions using *in situ* medium energy ion scattering analysis. *J. Vac. Sci. Technol. A*, **25**, 141–147.

Coondoo I., Agarwal S. K. & Jha A. K. (2009) Ferroelectric and piezoelectric properties of tungsten substituted $SrBi_2Ta_2O_9$ ferroelectric ceramics. *Mater. Res. Bull.*, **44**, 1288–1292.

Copel M. (2008) Reaction of barium oxide threshold voltage tuning layers with SiO_2 and HfO_2/SiO_2 gate dielectrics. *Appl. Phys. Lett.*, **92**, 152909.

Delabie A., Puurunen R. L., Brijs B., Caymax M., Conard T., Onsia B., Richard O., Vandervorst W., Zhao C., Heyns M. M., Meuris M., Viitanen M. M., Brongersma H. H. & de Ridder M. (2005) Atomic layer deposition of hafnium oxide on germanium substrates. *J. Appl. Phys.*, **97**, 064104.

Demkov A. A. & Navrotsky A. (2005) *Materials Fundamentals of Gate Dielectrics*, Dordrecht, Springer.

Doyle B. L. & Peercy P. S. (1979) Technique for profiling ^1H with 2.5 MeV Van de Graaff accelerators. *Appl. Phys. Lett.*, **34**, 811–813.

Frank M. M., Kim S., Brown S. L., Bruley J., Copel M., Hopstaken M., Chudzik M. & Narayanan V. (2009) Scaling the MOSFET gate dielectric: From high-k to higher-k. *Microelectr. Eng.*, **86**, 1603–1608.

Goncharova L. V., Starodub D. G., Garfunkel E., Gustafsson T., Vaithyanathan V., Lettieri J. & Schlom D. G. (2006) Interface structure and thermal stability of epitaxial $SrTiO_3$ thin films on Si(001). *J. Appl. Phys.*, **100**, 014912.

Goncharova L. V., Dalponte M., Gustafsson T., Celik O., Garfunkel E., Lysaght P. S. & Bersuker G. (2007) Metal-gate-induced reduction of the interfacial layer in Hf oxide gate stacks. *J. Vacuum Sci. Technol. A*, **25**, 261–268.

Gustafsson T. (2009) Medium energy ion scattering for near surface structure and depth profiling. In Hellborg R., Whitlow H. & Zhang Y. (Eds.) *Ion Beams in Nanoscience and Technology*. Berlin, Springer.

Gustafsson T., Garfunkel E., Goncharova L. V., Starodub D. G., Barnes R., Dalponte M., Bersuker G., Foran B., Lysaght P. S., Schlom D. G., Vaithyanathan V., Hong M. & Kwo J. (2006) Structure, composition and order at interfaces of crystalline oxides and other high-κ materials on silicon. In Gusev E. (Ed.) *Defects in High-κ gate dielectric stacks*. St. Petersburg, Elsevier.

Hellborg R. (2009) Ion accelerators for nanoscience. In Hellborg R., Whitlow H. & Zhnag Y. (Eds.) *Ion Beams in Nanoscience and Technology*. Berlin, Springer.

Im J., Krauss A. R., Lin Y., Schultz J. A., Auciello O., Gruen D. M. & Chang R. P. H. (1996) *In situ* analysis of thin film deposition process using time of flight (TOF) ion beam analysis methods. *Nucl. Instrum. Meth. B*, **118**, 772–781.

Kim H., McIntyre P. C., Chui C. O., Saraswat C. & Stemmer S. (2004) Engineering chemically abrupt high-κ metal oxide/silicon interfaces using an oxygen-gettering metal overlayer. *J. Appl. Phys.*, **96**, 3467–3472.

Kim H. S., Ok I., Zhang M., Lee T., Zhu F. & Yu L. L., J.C (2006) Metal gate-HfO_2

metal-oxide-semiconductor capacitors on n-GaAs substrate with silicon/germanium interfacial passivation layers. *Appl. Phys. Lett.*, **89**, 222903.

Krauss A. R., Lin Y., Auciello O., Lamich G. J., Gruen D. M., Schultz J. A. & Chang R. P. H. (1994) Pulsed ion beam surface analysis as a means of *in-situ* real-time analysis of thin films during growth. *J. Vac. Sci. Technol. A*, **12**, 1943–1957.

Krauss A. R., Im J., Schultz J. A., Smentkowski V. S., Waters K., Zuiker C. D., Gruen D. M. & Chang R. P. H. (1995) *In situ* analysis of thin film deposition process using time of flight (TOF) ion beam analysis methods. *Thin Solid Films*, **270**, 130–136.

Krauss A. R., Im J., Smentkowski V. S., Schultz J. A., Auciello O., Gruen D. M., Holocek J. & Chang R. P. H. (1998) Ion beam deposition and surface characterization of thin multi-component oxide films during growth. *Mater. Sci. Eng.*, **A253**, 221–233.

Lee H. D., Feng T., Yu L., Mastrogiovanni D., Wan A., Gustafsson T. & Garfunkel E. (2009) Reduction of native oxides on GaAs during atomic layer growth of Al_2O_3. *Appl. Phys. Lett.*, **94**, 22108.

Lee H. D., Feng T., Yu L., Mastrogiovanni D., Wan A., Garfunkel E. & Gustafsson T. (2010) ALD growth of Al_2O_3 on GaAs: oxide reduction, interface structure and CV performance. *Phys. Stat. Sol. C: Curr. Topics Solid State*, **7**, 260–263.

Marchut L., Buck T. M., Wheatley G. H. & McMahon C. J. (1984) Surface-structure analysis using low-energy ion-scattering. *Surf. Sci.*, **141**, 549–566.

Peterson J. J., Young C. D., Barnett J., Gopalan S., Gutt J., Lee C.-H., Li H.-J., Hou T.-H., Kim Y., Lim C., Chaudbary N., Moumen N., Lee B.-H., Bersuker G., Brown G. A., Zeitzoff P. M., Gardner M. I., Murto R. W. & Huff H. R. (2004) Subnanometer scaling of HfO_2/metal electrode gate stacks. *Electrochem. Solid State Lett.*, **7**, G164–G167.

Poschenrieder W. P. & Oetjen G.-H. (1972) New directional and energy focusing time of flight mass spectrometer for special tasks in vacuum and surface physics *J. Vac. Sci. Technol.*, **9**, 212.

Preisler E. J., Guha S., Copel M., Bojarczuk N. A., Reuter M. C. & Gusev E. (2004) Interfacial oxide formation from intrinsic oxygen in W–HfO_2 gated silicon field-effect transistors. *Appl. Phys. Lett.*, **85**, 6230–6232.

Reichelt T., Dollinger G., Bergmaier A., Datzmann G., Hauptner A. & Korner H. (2002) Sensitive 3D hydrogen microscopy by proton–proton scattering. *Nucl. Instr. Meth. B*, **197**, 134–149.

Schaeffer J. K., Fonseca L. R. C., Samavedam S. B., Liang Y., Tobin P. J. & White B. E. (2004) Contributions to the effective work function of platinum on hafnium dioxide. *Appl. Phys. Lett.*, **85**, 1826–1828.

Schindler G., Hartner W., Joshi V., Solayappan N., Derbenwick G. & Mazure C. (1997) Influence of Ti-content in the bottom electrodes on the ferroelectric properties of $SrBi_2Ta_2O_9$ (SBT). *Integrated Ferroelectrics*, **17**, 421–432.

Schlom D. G., Guha S. & Datta S. (2008) Gate oxides beyond SiO_2. *MRS Bull.*, **33**, 1017–1025.

Sreenivas K., Reaney I., Maeder T., Setter N., Jagadish C. & Elliman R. G. (1994) Investigation of Pt/Ti bilayer metallization on silicon for ferroelectric thin film integration. *J. Appl. Phys.*, **75**, 232–239.

Taglauer E., Englert W., Heiland W. & Jackson D. P. (1980) Scattering of low-energy ions from clean surfaces – comparison of alkali-ion and rare-gas-ion scattering. *Phys. Rev. Lett.*, **45**, 740–743.

Tenne D. A., Turner P., Schmidt J. D., Biegalski M., Li Y. L., Chen L. Q., Soukiassian A., Trolier-McKinstry S., Schlom D. G., Xi X. X., Fong D. D., Stephenson G. B., Thompson C. & Streiffer S. K. (2009) Ferroelectricity in ultrathin $BaTiO_3$ films:

probing the size effect by ultraviolet Raman spectroscopy. *Phys. Rev. Lett.*, **103**, 177601.

Tromp R. M., Kersten H. H., Granneman E., Saris F. W., Koudijs R. & Kilsdonk W. J. (1984) A new UHV system for channeling blocking analysis of solid surfaces and interfaces. *Nucl. Instr. Meth. B*, **4**, 155–166.

Turkenburg W. C., Soszka W., Saris F. W., Kersten H. H. & Colenbrander B. G. (1976) Surface-structure analysis by means of Rutherford scattering methods to study surface relaxations. *Nucl. Instr. Meth. B*, **132**, 587–602.

van der Veen J. F. (1985) Ion beam crystallography of surfaces and interfaces. *Surf. Sci. Rep.*, **5**, 199.

Wang Y., Ho M.-T., Goncharova L. V., Wielunski L. S., Rivillon-Amy S., Chabal Y. J., Gustafsson T., Moumen N. & Boleslawski M. (2007) Characterization of ultra-thin hafnium oxide films grown on silicon by atomic layer deposition using tetrakis(ethylmethyl-amino) hafnium and water precursors. *Chem. Mater.*, **19**, 3127–3138.

Wu L., Solovyov V. F., Wiesmann H. J., Zhu Y. & Suenaga M. (2002) Mechanisms for hetero-epitaxial nucleation of YBa_2Cu_3O similar to 6.1 at the buried precursor/$SrTiO_3$ interface in the postdeposition reaction process. *Appl. Phys. Lett.*, **80**, 419–421.

Zurbuchen M. A., Lettieri J., Fulk S. J., Jia Y., Carim A. H., Schlom D. G. & Streiffer S. K. (2003) Bismuth volatility effects on the perfection of $SrBi_2Nb_2O_9$ and $SrBi_2Ta_2O_9$ films. *Appl. Phys. Lett.*, **82**, 4711–4713.

7
Spectroscopies combined with reflection high-energy electron diffraction (RHEED) for real-time *in situ* surface monitoring of thin film growth

P. G. STAIB, Staib Instruments, Inc., USA

Abstract: The electron beam used for reflection high-energy electron diffraction (RHEED) can also be used for various surface analysis techniques such as X-ray emission spectroscopy (XES), total reflection angle X-ray spectroscopy (TRAXS), Auger electron spectroscopy (AES), cathodoluminescence (CL), and reflection electron energy loss spectroscopy (REELS). These techniques provide unique ways to perform *in situ*, during growth, monitoring of the surface atomic composition and chemical state. The basic electron–surface interactions and emission processes are described and the experimental techniques are presented. Typical examples of applications illustrate the capabilities of each technique and are compared in terms of information and experimental complexity with a special emphasis on their compatibility with large vacuum chambers and their challenging operating conditions.

Keywords: *in situ* surface analysis, Auger electron spectroscopy (AES), reflection electron energy loss spectroscopy (REELS), X-ray emission spectroscopy (XES), total reflection angle X-ray spectroscopy (TRAXS), cathodoluminescence (CL).

7.1 Introduction

In addition to reflection high-energy electron diffraction (RHEED) observations, the high-energy electron beam offers unique capabilities for using several spectroscopic methods to provide valuable information about the elemental composition and the chemical environment of the growing surface. The techniques used presently are X-ray fluorescence spectroscopy (XRF) with a special, more surface-sensitive application, total refraction angle X-ray spectroscopy (TRAXS), Auger electron spectroscopy (AES), reflection electron energy loss spectroscopy (REELS), and, more limited to band structure analyses, cathodoluminescence (CL) spectroscopy. The aim of these applications is to gather additional information about the chemical composition of the surface layers. The primary interest is to measure *in situ* and during growth the composition of the material as given by the ratio of atomic concentrations. The composition of the surface itself may be different from that of deeper layers, especially at lower growth temperatures. The actual chemical composition of the surface during growth and how it correlates to the composition and quality of the grown material are also of interest and are especially important when using gas sources in order to find and to understand the adjustment of material flux and sample temperature.

The feasibility of *in situ* spectroscopy combined with RHEED in a growth chamber has a long history. Even in the 1970s, in the early stage of molecular beam epitoxy (MBE), X-ray detectors were added to growth chambers and provided unique material information. AES and X-ray photoelectron spectroscopy (XPS) were also implemented, giving more surface-specific information. Many combined devices have since been used, but these techniques remained isolated attempts and are not yet part of the standard basic tools to control the growth process.

The scope of this chapter is to describe the present status of the available spectroscopies, their potential developments in light of the information provided, and the development of new dedicated instruments. This chapter is restricted to spectroscopies combined with RHEED and used *in situ* and operated in *real time* during the growth process, in contrast to the experimental devices requiring an *in situ* sample transfer for analysis or requiring the interruption of the growth process during data acquisition. The most important need of *in situ* analyses is the quantitative measurement of atom concentration on and near the surface. When successful, these techniques are likely to become major tools for controlling the growth of compound materials.

Section 7.2 presents a short overview of the physical parameters involved for understanding the measured signals in terms of atomic densities and sample geometry. Section 7.3 is an overview of the present experimental set-ups and results. Section 7.4 compares results obtained by each technique and highlights developments expected in the near future.

7.2 Overview of processes and excitations by primary electrons in the surface

Fast primary electrons (PE) from the RHEED beam penetrate the surface and generate cascades of events in the near-surface region. These interactions or collisions are classified into elastic and inelastic scattering processes. Each process has a specific cross-section and an associated specific mean free path (MFP), corresponding to the average distance traveled between successive specific interactions. Most electrons penetrate the solid and undergo a large number of elastic scattering events with a wide range of deflection angles, leading to a zigzag path inside the crystal lattice. Figure 7.1(a) shows the impact of PE at normal incidence and Fig. 7.1(b) at grazing incidence. Many

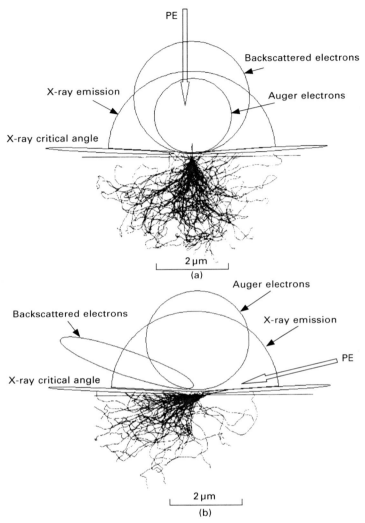

7.1 Trajectories of fast primary electrons inside the surface as calculated using a Monte Carlo computer simulation for (a) normal incidence angle and (b) grazing incidence angle. The angular distributions of backscattered electrons (BSE), Auger electrons, and X-rays emitted from the surface are also depicted.

interactions are inelastic events that progressively reduce the kinetic energy of the PE until it occupies an energy level of the crystal band structure. Some trajectories will eventually bring electrons back toward the surface. They may leave the surface and be emitted with a wide distribution of kinetic energies, ranging up to the primary energy, and a wide angular distribution. They build the flux of backscattered electrons (BSE) leaving the surface. This BSE flux also interacts with the solid and increases the efficiency of the PE flux by creating additional excitations in the surface region. The BSE electron flux represents a major correction factor for quantification.

7.2.1 Elastic scattering and elastic mean free path (EMFP)

Elastic scattering is a process associated with large scattering angles with (almost) no energy loss (Murata, 1974; Reimer, 1985, p57). The elastic mean free path (EMFP) is the average distance traveled by the PE between two elastic scattering events and measures the strength of process. EMFP values for most elements and compounds have been calculated and are available as databases. Figure 7.2 shows the EMFP for silicon as function of the electron kinetic energy as tabulated from the database of Werner (2003, p235, 2010). Detailed understanding of the microscopic behavior of the PE is obtained using Monte Carlo computer simulations as shown in Fig. 7.1. The electron paths and spatial distribution of the electron near the surface becomes apparent and the distribution of the BSE flux can be accurately calculated, even at grazing incidence angles. Energy and angular distributions of BSE are in good agreement with experimental data (Ichimura and Shimizu, 1981; Ding *et al.*, 2006). A recent program CASINO by Drouin *et al.* (2007),

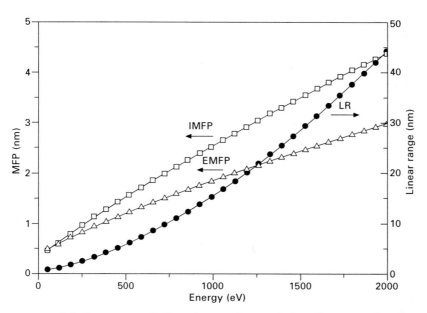

7.2 Linear range (LR) and mean free path for silicon as a function of the electron energy. The elastic (EMFP) and inelastic (IMFP) paths and range are given in nm.

available for download from the web, allows customized calculation of the BSE factors for user specific geometries and materials. The calculated electron trajectories show how the BSE flux is the result of multiple small angle scattering events rather than one single large angle scattering event, as is the case for Bragg diffraction. There is the formation of a shower-like excitation volume under the surface. PE kinetic energy is progressively dissipated inside the excitation volume until the electron becomes part of the valence or conduction band.

7.2.2 Primary electron range and penetration depth

The range of primary electrons of kinetic energy E_p is a measure for the maximum penetration depth of the beam. It can be described, at normal incidence, using a simple power formula (Reimer, 1985, p96)

$$R = a\, E_p^n \qquad [7.1]$$

where a and n are material specific constants. The range R is expressed in $\mu g\, cm^{-2}$. This relation is valid in an energy range 5–25 kV (Everhart and Hoff, 1971; Sogard, 1980). For Al or Si, the range is given by $R = 4.0\, E_p^{1.5}$ with E in kV. The range for Cu is given by $R = 9.0\, E_p^{1.5}$ showing that the energy variation of R is quite independent of the atomic number Z when measured as mass-thickness in units of $\mu g\, cm^{-2}$. This range corresponds to the maximum penetration of the PE beam at normal incidence. The range in units of cm is obtained by the density ρ in $g\, cm^{-3}$. More accurate calculations performed by Werner (2003) are available from the database (Werner, 2010). The range values, as shown in Fig. 7.2 for Si, are much larger than the EMFP or inelastic mean free path (IMFP).

7.2.3 Inelastic scattering processes and characteristic energy losses (CEL)

The PE can lose energy in several ways. The largest energy losses are the ionization of core energy levels and bremsstrahlung emission processes. Smaller loss values are related to plasmon and band excitation which are characteristic energy losses (CEL). Finally, small energy losses are due to phonon interactions and the creation of electron–hole pairs.

Inner shell ionization processes

The PE of energy E_p ejects an inner shell electron from an energy level E_i, leaving a vacancy in an inner shell energy level. The cross-section for this core level ionization process can be calculated using the formula given by Gryzinski (1965):

$$\sigma_i = 6.51 \times 10^{-14} \, N_i G(U_i)/E_i^2 \quad [7.2]$$

with $G(U_i)$ a universal function depending only on the overvoltage parameter $U_i = E_p/E_i$. N_i is the number of electrons in the ionized shell. K shells contribute with two electrons and M_{45} shells with six electrons. Examples of calculated cross-sections as a function of the PE energy are shown in Fig. 7.3. The two major characteristics of the distributions are:

- The maximum cross-section for ionization is reached for primary electron energies 3 to 6 times larger than the ionization energy. For solids, the position of the maximum ionization yield is shifted toward higher PE energies because of the backscattered electrons as discussed below.
- The value of the maximum cross-section is larger for lower ionization energies E_i, as shown in Fig. 7.3 in the case of silicon. The cross-section at $E_p = 15\,\text{keV}$ for the low-energy L_{23} level ($E_i = 104\,\text{eV}$, six electrons) is 100 times larger than for the K level (two electrons, $E_i = 1844\,\text{eV}$). The difference in the values of the Si cross-sections is the dominant factor explaining the large variations of the signal intensities for different transitions in AES and X-ray emission spectroscopy (XES). On the other hand, the variation of σ for L_{23} shell ionization shows a weaker dependence on Z for neighboring elements like Fe to As for PE energies above 10 keV.

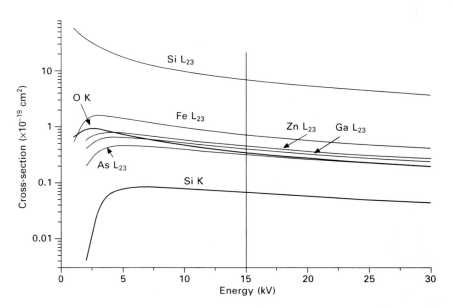

7.3 Ionization cross-sections for the K and L_{23} shell of selected elements as a function of the energy of the RHEED electron beam.

CEL: plasmons and band transitions

Electrons located in outer energy bands can be excited either as collective density oscillations known as plasmon excitation, or as single electron inter- and intra-band transitions. The result is the creation of CEL having a well-defined energy ΔE generally in the range from a few eV up to about 40 eV. The PE loses this amount of kinetic energy and is scattered by a small angle as required by the conservation of energy and momentum.

Plasmon losses correspond to the eigenmodes of resonance of the electron gas forming the outer electron shell, valence, and conduction bands. The plasmon frequency for a metal with an ideal free electron gas, like aluminum, is given by

$$\omega_p = (4\pi\, e^2/m)^{1/2}\, (n)^{1/2} \qquad [7.3]$$

with e the electron charge, m the effective mass of the electrons, and n the electron density of the conduction electrons (Raether, 1980; Ferguson, 1989). The excitation of plasmons is not restricted to free electron gas, but also exists for insulators and semiconductors because the plasmon energy, commonly in the range 5–30 eV, is larger than the electron binding or gap energy (Raether, 1980, p15). The plasmon loss energy is $\Delta E = \hbar\omega_p$ and is directly related to the density of state n and, therefore, to the chemical environment of the surface atoms. For instance, CEL of Al and Al_2O_3 are very different with plasmon loss energies of 15 and 23 eV respectively.

Plasmon excitations come in two forms, surface and volume plasmons. The surface plasmon $\omega_{p,s}$ corresponds to electron density fluctuations in the boundary surface and the loss energy is

$$\Delta E_{p,s} = \Delta E_{p,v}/\sqrt{2} \qquad [7.4]$$

for a free electron gas model. Plasmon losses are generally observed as multiple losses. The probability for an electron to suffer n plasmon losses depends on the ratio between the path length d traveled and the IMFP λ and is given by the Poisson distribution

$$Pn = 1/n!\, d/\lambda \qquad [7.5]$$

as represented in Fig. 7.4. The no-loss probability P_0 decreases very rapidly even for small values of d/λ. For $d = \lambda$, P_0 is reduced by about 37% and equals P_1, the first loss peak. Even a layer thickness of 0.5 nm will cause the no-loss peak to decrease to 60% of its value. This is a sensitive method for measuring the thickness of deposits like a gas adsorbed on the surface. The ratio between intensities of multiple plasmon losses can be compared to give an estimate of d/λ. A plasmon excitation can also occur during the ionization-recombination process (intrinsic excitation). An atomic layer adsorbed on the surface will suffer characteristic energy losses even if the escape distance traveled is negligible compared with the IMFP. This process

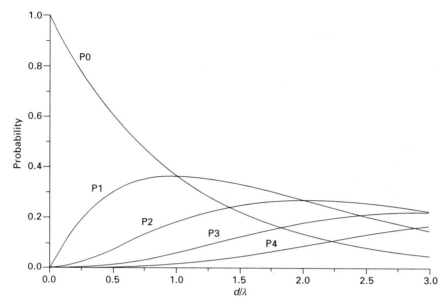

7.4 Probability for multiple energy loss as a function of the ratio between the path length d and the mean free path λ.

adds to the excitation of extrinsic losses occurring during travel through surface layers.

Band transitions are excitations of outer shell electrons. Ionization occurs when a valence or conduction electron is ejected, resulting in the creation of a vacancy. In contrast to optical absorption, the available momentum of the PE allows for direct and indirect band transitions. Band transition losses often occur in the same energy range as plasmons and overlap. The energy loss distribution is closely related to the optical properties of the surface. The energy loss function is given by the imaginary part of the complex dielectric constant (Raether, 1980, p35) and CEL distributions can be deduced from optical data. The strength of the energy losses is given by their IMFP which is the mean distance traveled by the electron between two CEL. The mean free path for a plasmon excitation is proportional to the electron kinetic energy, except in the low kinetic energy range (<50 eV) where it increases. The IMFP shown for Si in Fig. 7.2 has a value of about 4 nm at 1800 eV and only 0.6 nm at about 100 eV. This is a large difference in escape depth for Auger or photoelectrons and is a major factor in the calculation of atom densities.

Continuous X-ray emission (bremsstrahlung)

Fast PE elastically scattered by nuclei in the crystal lattice are subject to strong acceleration and can emit X-ray photons of energy ranging up to the

PE energy (Reimer, 1985, p158). The probability for this process is small and the process does not contribute significantly to the stopping power, but it generates a significant background continuum of X-rays. This background distribution adds to the characteristic X-ray lines and lowers the signal to noise ratio and the detection limit. The production rate Iv is given by Small et al. (1987)

$$Iv = 10^5 \times e^B [Z(U-1)]^M \qquad [7.6]$$

with e the natural log base, Z the atomic number, and $U = E_p/E_i$ the overvoltage parameter. The constants are $M = 1.05$ and $B = 5.80$ in the PE range 5 to 40 kV. The contribution increases almost linearly with Z and the detection of a low-Z element deposited on a high-Z substrate is more difficult than the other way around.

7.3 Recombination and emission processes

Core level vacancies created by fast PE impact are filled by electrons from higher shells and the energy is released either as X-ray photons or Auger electrons. The probability for the X-ray emission rather than an Auger electron emission is given by the fluorescence yield factor ω. Figure 7.5 shows the fluorescence yield for different shells plotted against the atomic number Z using data from Krause (1979) and Segre (MUCAL). The X-ray photon energy ranges up to several 10 keV, but the kinetic energy range of Auger electrons is much narrower. A range up to 2500 eV is sufficient to include the main Auger lines from all elements. The K shell Auger lines extend up to $Z = 16$ (S), the L shells up to $Z = 46$ (Rh), the M and N shells up to $Z = 85$ (At). These ranges are indicated in Fig. 7.5 and show that the fluorescence yields for the K, L, M shells are correspondingly low. For instance, the fluorescence yield for oxygen K is 0.83% and for silicon K is 5%. The fluorescence yields are small for low-energy transitions and Auger electron emission is the dominating process. Theoretical treatment of the Auger process is complex because of the existence of relaxation effects due to coulomb screening of the double ionized atom and possible interatomic cross-transitions (Coster–Kronig transitions). General descriptions of these emission processes are given in Reimer (1985) and Ferguson (1989) and are more detailed for XPS and AES applications in Briggs and Grant (2003).

Auger and characteristic X-ray lines generally have a multiplet structure. This structure can be resolved with a suitable energy resolution of the detector system. This is normally the case for AES energy analyzers and X-ray dispersive (XDS) spectrometers, with resolution in the range 1–5 eV. Energy dispersive X-ray detectors (EDS) have limited energy resolution, about 100–150 eV, and many fine structures will appear as a single peak.

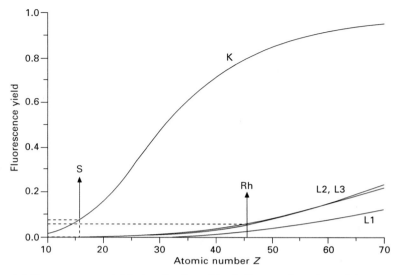

7.5 Fluorescence yields for the K and L shells as a function of the atomic number Z. The ranges corresponding to the KLL and LMM Auger transitions are marked and correspond to low fluorescence yield. After M.O. Krause and C. Segre.

7.3.1 Quantification of the signal intensities

The emission yield is primarily given by the cross-section for ionization, the escape depth, and the backscattering factor. The major experimental difference between XRF, AES, and REELS is the large difference in escape depth of photons or electrons. X-ray photons can escape from several micrometers below the surface whereas AE will be limited to a few nanometers. The absorption of X-rays is dominated by photo-ionization (mostly self-absorption) and Compton scattering. The resulting MFP value is in the range of several micrometers. In practice, the X-ray escape depth can be larger than the penetration of the PE beam for an energy of 10 keV. Most of the emitted X-ray photons generated near the surface are able to escape into vacuum and have a nearly isotropic angular distribution as shown in Fig. 7.1. The surface sensitivity of XES can be improved using grazing takeoff angles as discussed below for TRAXS. In contrast, the IMFP for Auger electrons is in the nanometer range (Ferguson, 1989, p25; Kanter, 1970) and the angular distribution more closely follows Lambert's cosine law; see Fig. 7.1.

A convenient way to quantify the data is to use the ratio method based on reference spectra. Figure 7.6 shows the basic experimental conditions found in growth experiments. In the first case, Fig. 7.6(b), two elements A and B with atomic numbers Z_A and Z_B are forming an alloy A + B. The reference spectra from each pure element, corresponding to an atomic concentration

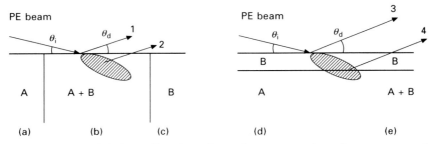

7.6 Models for calibration of atomic concentrations for a compound A + B (a, b, c), for a deposited layer A on B (d) and for a layer of B on A + B (e).

of 100%, is measured and gives the signal intensities $I_A(100)$ and $I_B(100)$. An alloy of concentration x of A and $1 - x$ of B gives the signal intensities $I_A(x)$ and $I_B(1 - x)$. The signal intensity for the pure element A of volume density N_A is given by

$$I_A^x = N_A^x I_p \sigma_A M_A D_A \qquad [7.7]$$

with σ_A the ionization cross-section for A for a PE beam of energy E_p and intensity I_p. M_A is a correction for matrix effects that include multiple effects such as the backscattering coefficient, the fluorescence yield, and the escape probability of the particle to be emitted from the surface. The escape probability depends on the IMFP for Auger and REELS electrons or on the absorption for X-rays. It also depends on the angle of detection θ_d. The instrumental factor D_A is the detection efficiency of the spectrometer for element A and includes the acceptance angle and transmission of the spectrometer. The detailed calculation of each correcting factor is complex and it can be simplified using reference data from pure elements. The ratio between the signals of an alloy of concentration x and the pure element is

$$I_A(x)/I_A(100) = N_A(x)/N_A(100) \cdot M_A/M_{A(x), B(1-x)} \qquad [7.8]$$

The correcting factor is now the ratio between the matrix factor for the alloy $M_{A(x), B(1-x)}$ to the pure element M_A. The major variations are those of the attenuation length and backscattering factors, and the correcting factor can be expressed for a signal from A as

$$d_{A(x), B(x)}/[d_A(100) \cdot B_A(100)] \qquad [7.9]$$

where d_A is the escape depth of a signal from A at concentration x and 100% and B_A is the backscattering factor for the same.

The second example is shown in Fig. 7.6(d) where element B is deposited on top of substrate A. Signal I_B from element B will grow linearly at the beginning of the growth until the surface coverage reaches 1 ML (monolayer). For a thicker layer, I_B will reach a steady value for deposits thicker than the escape depth of signal B in layer B. Concurrently, the signal A will decrease as the signal I_A has to cross the layer B and finally will vanish. The absorption is characterized by the IMPF of I_A through layer B and by the detector take-off angle θ_d. Signal I_A will decrease as

$$I_A(d) = I_A(0) \exp - d/[\cos(\theta_d) \lambda] \quad [7.10]$$

with d the thickness (coverage) of B and λ the IMFP at the energy of line A in element B.

A frequent case is when growing a compound using a larger flux of one element, like As in GaAs or like O and O_2 in oxide growth, where one element may accumulate on the surface. Figure 7.6(e) shows the case of the growth of element B on A + B where element B builds up on the surface. The signal I_B comes from two contributions, one from the bulk and the other from the surface layer. I_B will grow as long as the surface coverage remains low, below 1 ML, because the attenuation in layer B is small and both contributions can add. The signal from A will not change much. When the layer thickness is larger than the attenuation length λ, signal I_B will saturate to its value for pure B material and signal A will vanish. Interestingly, in the case that the thickness is near λ and that the atomic number Z of A is larger than that of B (as for instance an oxygen layer on a metal oxide substrate), the I_B signal will show a peak because the flux of BSE from A + B into the surface layer B is larger than the BSE flux from pure B. This example shows that the deduction of atomic densities from signal strengths requires the proper knowledge and modeling of the growth conditions on the surface.

7.4 Descriptions and results of *in situ* spectroscopies combined with reflection high-energy electron diffraction (RHEED)

This review focuses on recent combined techniques implemented *in situ* with RHEED and able to deliver results during the growth process, excluding analytical instruments built *in situ* but requiring sample transfer or interruption of the deposition process for acquisition of data. The environment of a growth chamber puts specific constraints on the instrument design in order to operate *in situ* and during growth.

- A large working distance between the sample and the detector system is required in order not to impair the atomic fluxes from the deposition sources.
- A good detection sensitivity allowing real-time data acquisition, fast enough to follow the growth process with acquisition times in the range 1–30 s is necessary for most processes.

- An outstanding resistance against material deposition is needed to ensure long-term stable operation over several months.
- The capability to operate in a wide range of pressures, from ultra-high vacuum (UHV) up to millitorrs, in order to be compatible with the operation of gaseous sources, is important.

Standard instruments are not directly suitable and must be modified to fit these applications. New instruments, specially designed for operation in growth chambers, are now being tested and will add new capabilities. The techniques presented here are CL, XRF, TRAXS, REELS, and AES.

7.4.1 *In situ* CL spectroscopy

CL is similar to photoluminescence (PL) spectroscopy because the recombination process is the same for both. There is, however, a major difference in the excitation process because, in the case of PL excited by a laser, the photon energy is fully absorbed, whereas electron excitation leads to a wide distribution of energy transferred. Further, the cross-sections for photon absorption have a peak at the transition threshold, but for electrons the increase is smooth near the threshold. PL can selectively excite specific transitions and CL will excite a wide range of transitions. The recombination process involves band transitions between the valence and conduction bands (Reimer, 1985, p289). Electrons from the valence band are excited into unoccupied states of the conduction band. A cascade of non-radiative phonon and electron excitations reduces their energy until they reach the bottom of the conduction band. The luminescence decay processes involve the creation of electron–hole pairs and excitons (Lightowlers, 1990). The transition can be direct or an indirect transition involving phonons in order to insure the conservation of momentum. The spatial extent of the region producing the CL signal is very large because all of the PE, BSE, and even most of the secondary electrons have sufficient energy to excite interband transitions and generate CL emission. The spatial area covers the full range of the cascade shown in Fig. 7.1. An additional broadening, due to the diffusion of the carriers during their lifetime before decay, further extends the emission volume (Reimer, 1985, p292). Therefore, the CL method basically delivers bulk information, except when observed transitions involve surface states.

Experimental set-up for CL spectroscopy

The CL set-up was developed for use in electron microscopes and involves the imaging of the beam spot through a large aperture optical collector (Reimer, 1985, p210). CL designs must be modified for *in situ* applications to accept larger sample size and to increase the clearance required between the sample and the optical system. In practice, only a limited beam aperture is focused onto the detector. In contrast with electron microscope chambers that are extremely dark, growth chambers contain various sources of stray light and in addition the sample may radiate when heated during the growth. Optical filters, adapted to the emitted CL spectral range can filter out the parasite light outside the CL emission range. A more efficient suppression method is the modulation of the PE beam intensity, using beam blanking, and measuring the signal using a lock-in amplifier detection technique (Lee and Myers, 2007).

Application of CL to the measurement of the substrate temperature

GaN having a direct band gap is a good CL emitter and the spectral distribution shows a peak energy corresponding to the band gap energy (Lee and Myers, 2007). The position of the maximum is very temperature sensitive. The peak shifts toward lower energy as the sample temperature increases. The temperature measured by CL represents the actual temperature of the very surface layer and represents a way to calibrate sample temperature between different chambers. Once calibrated, CL is able to detect variations of the surface temperature more precisely than the usual thermocouple measurement. CL provides a way to cross-calibrate temperature measured in different GaN growth chambers.

7.4.2 In situ XES combined with RHEED

XRF, also XES, is performed *in situ* using different detectors. One of the first experiments combining RHEED and XRF used a wavelength dispersive spectrometer (Sewell and Cohen, 1967). The wavelength dispersive spectrometer (WDS) is a windowless device and can detect low-energy X-ray lines, such as oxygen Kα. The energy resolution is sufficient to resolve the multiplet structure of the lines, but the major disadvantage is that the detected signal is low and the acquisition requires a long measuring time. More recent experiments use the EDS detector with the advantage of detecting all incoming X-ray photons and delivering a signal pulse with an amplitude proportional to the photon energy. Integration times are much shorter and compatible with the speed of growth of the surface. The energy resolution of the detector is, however, not as good as for the WDS system and lines of neighboring Z elements often strongly overlap.

Experimental set-up for RHEED-XRF in situ monitoring

The angular distribution of characteristic X-rays is basically isotropic (see Fig. 7.1), except in the range of very grazing angle of emission where the refraction effects become dominant. EDS detectors have a large sensitive area and accept radiation over a large solid angle. The acceptance solid angle must be reduced in order to block X-rays emitted from the chamber walls by X-ray fluorescence or BSE impact. A collimator system consisting of successive apertures is inserted between sample and detector. The Si(Li) detector crystal must be cooled in order to reduce the thermal background noise level. Cooling can use liquid nitrogen contained in a Dewar tank or, more simply, stacked Peltier cooling stages. Most detectors are sealed with a Be window in order to keep the crystal free of moisture condensation and contamination. In addition, the foil removes and blocks the light and BSE. Unfortunately, the Be window also acts as a filter absorbing the lower energy part of X-ray photons. The sensitivity of the detector is progressively reduced for photon energies below about 1 keV, cutting off, for instance, the oxygen Kα radiation. The Be foil can be replaced with lower Z materials, such as sapphire and C (in the form of polymer materials), or totally removed by carefully keeping the detector under controlled vacuum conditions. Because low Z window materials do not efficiently block fast BSE, a strong permanent magnetic can be used to dump the electrons before reaching the detector.

An additional requirement is to keep the detector crystal and the window foil free of material deposition. This is very important for quantitative analyses because any deposit will cause a nonlinear, selective absorption of X-rays and will selectively modify the intensities of measured peaks. A solution for materials with a low evaporation temperature is to use a heat control able to keep the foil clean during the process. This technique is used for elimination of the deposition of As during GaAs/AlGaAs by Pellegrino *et al.* (1998). Another approach is the use of an easily replaceable thin Be foil (Sun *et al.*, 2009). In case the detector head cannot be easily accessed, moving a film like Mylar in front of the detector is another possible approach. Finally, X-ray energy dispersive detectors are not bakeable and must be placed far enough from the chamber wall or mounted on a mechanical retraction to move the detector far enough to stay at ambient temperature.

Quantification of characteristic X-ray line spectra

The signal intensity $I_A(x)$ of an element A is given by general Formula 7.7. The matrix factor is given by the product

$$M = \omega_i \cdot d/\cos(\theta_d) \, B \cdot F \qquad [7.11]$$

of the fluorescence yield ω_i for transition i, the backscattering factor B, the fluorescence factor F, and the attenuation length, corrected for the takeoff angle θ_d. The factor B becomes more significant when both incident and take off angles are small and is calculated by Monte Carlo simulations (CASINO). The standard method for quantitative calculation of atomic concentrations is described by Reimer (1985, p365) and referred to as the ZAF procedure. The method is based on the use of reference spectra from standards and followed by a linear combination of the standards to fit the measured signals. The accuracy of this simple method is much improved by introducing a correction factor for the Z dependence. It is assumed that the concentration is proportional to the ratio between the measured and the standard signal (corresponding to 100% concentration). Then the correction due to matrix effects, the ZAF factor, is introduced, taking into account variations of the BSE backscattering factor for different Z, of the absorption coefficient, and of the fluorescence factors between the actual sample and the standards. EDS-XRF combined with RHEED was used by Pellegrino *et al.* (1998) for quantitative monitoring of the surface composition of epitaxially grown InGaAs on GaAs. The stability of the signal over a long period of time is very reproducible and the accuracy of the mole fraction of In is <0.1% after background correction. Quantitative analyses are performed by fitting Gaussian peaks for each element. The full width half maximum (FWHM) of the distribution is known to be the energy resolution of the detector and deconvolution of multiple, overlapping lines is therefore accurate and reliable (Hashimoto *et al.*, 2009).

The surface sensitivity of EDS-XRF was tested by Ino *et al.* (1980) for the deposition of thin Ag layers on Si(111) surface. The detection limit is very good, less than 1% of a monolayer. The measurement of a high-Z material deposited on low-Z substrate is favorable for obtaining the highest detection limit because the bremsstrahlung background from the substrate is low and the PE backscattering and fluorescence effects are minimized.

7.4.3 Increasing the surface sensitivity with RHEED-TRAXS

The signal in XRF is dominated by the bulk emission and has a limited surface sensitivity. When the take-off angle of the X-ray signal is limited to a small angular range very close to the surface plane, the X-ray signal becomes highly surface sensitive (Hasegawa *et al.*, 1985). TRAXS is achieved when the geometrical conditions in Fig. 7.7 are fulfilled (Wang, 1996, p356). The X-rays emitted from a surface atom A at decreasing angle with the surface, labeled (1) and (2), are weakly refracted. The refractive index for X-rays is slightly smaller than unity and the emerging angle is larger than the incidence angle. X-rays emitted from A parallel to the surface (3) are refracted into vacuum at an angle θ_c. This angle is the critical angle for total refraction in XRF. Reversing the beam direction, X-rays with an incidence angle equal or smaller than θ_c will be totally reflected from the surface. There is no radiation emerging from the surface layer in the angular range between θ_c and the surface plane. The signal from deeper atomic layers B will not be

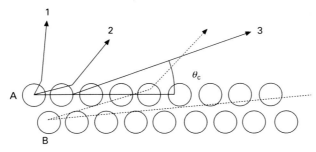

7.7 Beam geometry for critical angle X-ray spectroscopy (TRAXS).

emitted under the critical angle because the path to the surface is large and they will be absorbed. As absorption will cause fluorescence, a part of the signal from deeper layers may be re-emitted by the surface atoms, a process similar to the backscattering of electrons. The high surface sensitivity of this technique relies first on the small angular range selected by the detector close to the total reflection angle suppressing contributions from larger angles (and deeper layers) and second on the strong absorption of X-rays emitted at the critical angle but originating from deeper atomic layers. The depth of information is related to the decay length of the evanescent wave and is estimated in the order of 2 to 3 nm (Hasegawa et al., 1985).

Measurement of the critical angle in RHEED-TRAXS

The values for the critical angle are in the milliradian range and afford a high mechanical stability and accuracy of the positioning of the collimator unit. In principle, the take-off angle must be adjusted for each line when working under critical angle conditions. The critical angle for a compound is given by the simplified formula

$$\theta_c = 7.35 \times 10^{-5} \lambda \, (\rho \, \Sigma Z / \Sigma M)^{1/2} \qquad [7.12]$$

for an X-ray of wavelength λ and a material of density ρ, mass number M, and atomic number Z. The critical angle depends both on the X-ray line energy and on the composition of the matrix. The angle is larger for lower X-ray energies (longer wavelength), for instance $\theta_c = 1.77°$ for Mg Kα (1254 eV) in MgO and only $\theta_c = 0.33°$ for Fe Kα (6403 eV) in BaFe$_{12}$O$_{19}$ (Sun et al., 2009). Accurate measurements of θ_c are given by Chandril et al. (2009) and Tompkins et al. (2006).

Measurement of layer thickness using attenuation by RHEED-TRAXS and in situ XPS

A different way to determine the thickness of a film during deposition is to measure the attenuation of a signal originating from the substrate. An example is given by Sun *et al.* (2009) for the deposition of MgO onto SiC. The relative intensity changes were observed under fixed RHEED electron energy (12.5 keV). Monitoring substrate Si Kα intensity decrease with film growth provides a real-time film thickness measurement. The film thickness is calibrated using *in situ* XPS measurements of the attenuation of the Si 2p3 photoelectron line at different angular incidence (angular resolved XPS).

Measurement of multilayer thickness using TRAXS and X-ray reflectivity

Chandrill *et al.* (2009) demonstrated that the angular dependence of XRF near the critical angle can be analyzed within the kinematic approach to yield useful structural information not only about a single thin film but also for a multilayered structure. Hence, it is possible to perform *in situ* quantitative structural characterization without the use of any reference samples. The X-ray reflectivity of a multilayer is measured *ex situ* using an X-ray beam of the same energy and scanned over the same angular range as the detected waves in TRAXS. This is like reversing the path of the X-ray waves and, according to the reciprocity theorem, the intensity of the X-rays emitted from a depth z inside a medium and observed at a grazing exit angle is proportional to the X-ray intensity, which is impinging at the same grazing incident angle and observed at the depth z inside the medium. Angular intensity profiles measured from multilayers of single layers of Mn and Y as well as bilayers containing Y on Mn and Mn on Y allow the determination of the layered structure and of the smoothness of the interface surfaces.

7.4.4 *In situ* reflection energy loss spectroscopy

Electron energy loss spectroscopy (EELS) is a familiar technique used in transmission electron microscopes for measuring the ionization losses suffered by the PE beam (Disko *et al.*, 1992). The PE beam of high energy E_p (larger than 50 keV) is transmitted through a thin sample foil and is energy filtered by a high-energy resolution analyzer. The energy distribution of the transmitted electrons is measured and shows a strong direct transmitted peak at E_p of electrons transmitted with no energy loss. The intensity decreases very sharply with increasing loss energies and shows step-like structure corresponding to the ionization threshold energies E_i of core level electrons. As not only core level ionizations but also CEL losses like plasmons are possible, the observed step can be shifted by one or more plasmon energies depending on the IMFP for plasmon excitation and film thickness (Egerton, 1992, p41). The thickness of the sample is selected to optimize the ionization yield and to keep the plasmon losses to a minimum.

REELS is a similar technique that is used in reflection instead of transmission. There is a major difference between the transmission and reflection modes. In transmission and under optimal thickness conditions, the PE direct beam is straight and its angular extension is limited by the small analyzer acceptance angle (typically 50 milliradian). The ionization process leads to a smaller scattering angle, especially at larger PE kinetic energies, and is accepted by the filter whereas elastically scattered PE and BSE are rejected. In reflection mode and under RHEED conditions, the total scattering angle between the PE beam and detector is large. Elastic scattering collisions are needed to backscatter the PE beam and consequently it is mixed with the flux of BSE emerging from the surface. In contrast with the transmission mode, the ionization loss structures sit on top of a larger distribution of BSE (see Fig. 7.1). Unlike for the transmission mode, there is little latitude to optimize the length traveled by PE in the surface region in order to optimize the probability of excitation. The only factors which can be used are the incidence angle and the beam energy E_p. The signal to background ratio (S/B) is therefore not as good as in transmission mode.

Experimental set-up for REELS analyses

A modified TEM (transmission electron microscope) energy filter for energy losses spectroscopy is used in reflection mode by Nikzad *et al.* (1992). A small pencil of diffracted or diffused beam is selected and enters the analyzer. The beam is energy filtered by a stigmatic imaging magnetic sector followed by a set of magnifying projection lenses. The focal plane contains a range of energies which are collected simultaneously. Another analyzer for REELS is a device that combines energy filtering and imaging. The full diffraction diagram is accepted and, by means of a set of electron lenses, is energy filtered and projected onto a fluorescent screen similar to the usual RHEED screen (Staib *et al.*, 1999). The angular distribution is preserved and RHEED diagrams are energy filtered as shown in Fig. 7.8. The filtering progressively removes the inelastic BSE part and, at the cut off energy E_p, only the elastic diffracted spots remain and energy losses are filtered out. REELS distributions are achieved by measuring the intensity of a selected area of the diffraction diagram and lowering the energy filter voltage from above the elastic peak down to a few 100 eV loss energy. The signal is measured using a lock-in detection technique in order to improve the signal to noise ratio.

Examples of in situ REELS distributions

The surface sensitivity of REELS is demonstrated by Nikzad *et al.* (1992) for the deposition of Ge onto Si. The growth of the Ge L_{23} ionization edge is recorded at a PE energy of 30 keV. A 0.15 nm deposited layer is clearly detected after normalization of the spectra. Weaker structures can be detected

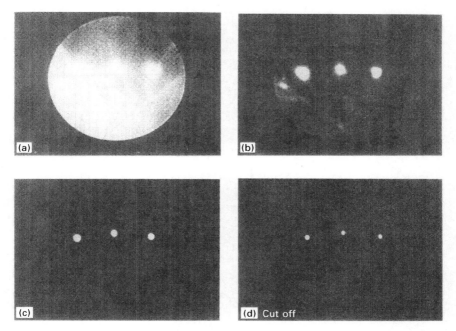

7.8 Energy filtered RHEED diagrams of (100) SrTiO$_3$ at filter voltages (a) 300 eV, (b) 100 eV, (c) 30 eV, and (d) 10 eV from the cut-off voltage.

after differentiation of the signal. A more detailed discussion of REELS in MBE is given by Wang (1996, p334).

Figure 7.9 shows the REELS distribution of SrTiO$_3$ recorded at E_p = 14.5 kV measured using the imaging energy filter described by Staib *et al.* (1999). The energy loss distribution includes the elastic peak and multiple CEL features. The losses are strongly apparent without background subtraction. The electron binding energy levels, available from Sevier (1972) and updated by Williams (2006), are used to identify the loss structures. The structures (6) correspond to Ti 3s (58.7 eV), (4) to Ti 3p1/2 3p3/2 (32.6 eV), (3) to Sr 4p1/2 (21.6 eV) and Sr 4p3/2 (20.1 eV). Loss energy (5) corresponds to O 2s (41.6 eV) but is shifted to about 47 eV as a result of a chemical shift.

Use of RHEED-REELS for monitoring the surface growth and roughness

The ratio between surface and volume plasmon is used to characterize, *in situ* during growth, the surface roughness of a deposited Al film on sapphire (Strawbridge *et al.*, 2006). The layer thickness was calibrated *ex situ* by Rutherford backscattering (RBS) and the surface roughness was measured by atomic force microscopy (AFM). The change of the energy

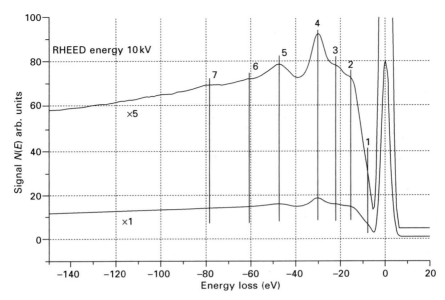

7.9 REELS of (100) SrTiO$_3$ near the elastic reflected peak measured at a PE energy of 14.5 keV.

loss distribution from the bare sapphire substrate to deposited Al layer is very sensitive to the Al atomic concentration. For thin layer deposits, only the surface plasmon is observed. For larger deposits in the range 3–14 nm, two behaviors are found; either the surface plasmon remains dominant correlating to a smooth surface deposit, or the volume plasmon dominate witnessing a rougher surface as confirmed by AFM scanning images. It is worth noticing that RHEELS data can be acquired in spite of the fact that sapphire is an excellent insulator. RHEELS is far less sensitive to charging effects than other analysis methods because the gain or loss in kinetic energy of the PE due to the surface charges compensates after backscattering. The PE enters and leaves the surface within a nanometer scale offset and the surface potential is homogeneous over this range. The same holds for the characteristic energy and ionization losses.

Monitoring the ionization loss for quantitative analyses

Quantitative analyses can be performed using the technique developed for transmission EELS as described by Leapman (1992). The intensity of the EEL structures is measured after background subtraction using a simple power law formula. The signal is then integrated over an energy width that includes the CEL region. A similar calculation can be used for reflection EELS, but an additional correcting factor is needed to account for backscattering effects

Spectroscopies combined with RHEED   201

(which are not present in transmission). The extraction of structures near the elastic peak, where ionization losses are superimposed on multiple CEL losses, requires a Gaussian background fitting.

7.4.5 *In situ* monitoring by Auger electron spectroscopy (RHEED-AES)

The escape depth of Auger electrons is given by the IMFP for a specific Auger line energy and material composition. A few relevant IMFP data are plotted in Fig. 7.10 as function of the kinetic energy using tabulated data from Powell (2000, NIST database 71). The plot shows that the IMFP depends on Z, but is even more dependent on the chemical state of the surface. Oxides like SiO_2 and ZnO have larger IMFPs than their pure components. The MFP is minimum for kinetic electron energies in the range 30–50 eV with values around 0.5 nm and increases to values 2–3 nm at 1000 eV (Ferguson, 1989, p25; Tanuma, 2003). When working with very thin deposited or adsorbed layers, the IMFP can alternatively be expressed in terms of monolayer coverage.

Experimental set-up for RHEED-AES

For *in situ* operation, the energy analyzer must fulfill the specific requirements listed previously. These constraints rule out the classical spectrometer designs

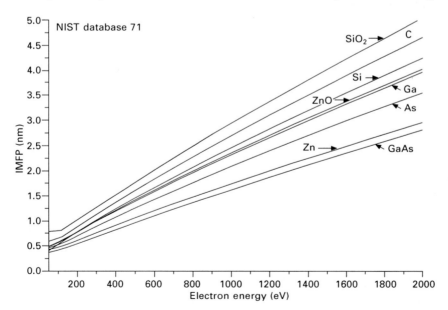

7.10 IMFP for several elements and compounds in the energy range of Auger electrons.

in favor of specially designed systems consisting of an electron optical lens system collecting the Auger electrons in front of the sample followed by an energy filter located further away from the sample, possibly behind the chamber wall. The angular position of the analyzer with respect to the surface is important. The analyzer should not be mounted at grazing take-off angle because the angular distribution of Auger electrons roughly follows Lambert's law of a cosine distribution. The maximum peak intensity is normal to the sample surface; see Fig. 7.1. In contrast, the BSE peak toward the direction of specular reflection. Thus, an Auger analyzer mounted with its axis normal to the sample will collect most of the Auger electrons, less BSE, and thus have the best sensitivity. The normal position has the advantage of not being sensitive to crystalline channeling effects affecting the angular distribution of Auger electrons, especially marked for low energy lines. The azimuthal rotation of the sample during acquisition is then possible. An additional electron gun, positioned far from grazing incidence angle (about 45°) and able to work over a wide energy range, is useful for calibration purpose in quantitative analyses and for measuring REELS distributions in a lower beam energy range.

In situ monitoring of oxygen during ZnO oxide growth on GaN substrates

ZnO is grown on top of a GaN substrate using a radio frequency (RF) plasma source for oxygen deposition. The growth process can be followed in real time using a fast Auger probe analyzer mounted at normal incidence angle using the experimental set-up given by Staib (2011). The Auger electron lines of Ga LMM (1066 eV), oxygen O KLL (503 eV) and Zn LMM (990 eV) are monitored. The growth rate is about 10 ML/min and at an acquisition speed of 6 V/s, the measuring time for an Auger line is 3–10 s. The Ga signal abruptly vanishes and the Zn signal grows rapidly when the Zn shutter opens; see Fig. 7.11(a). The oxygen line is present on the surface, provided by the residual gas pressure in the chamber. The oxygen K LVV line grows further on opening the oxygen source shutter (see Fig. 7.11b), and reaches a stable value. The normalized intensity ratio of O KLL to Zn LMM is a measure of the ratio of atomic concentrations.

In situ monitoring of growth of MgO and of $Cr_x Mo_{(1-x)}$ on MgO substates

Chambers *et al.* (1996) used RHEED combined with AES and *in situ* XPS to study the homoepitaxy growth of MgO on MgO(001) substrates. The sample temperature is kept at 750 °C during deposition and oxygen is provided by an electron cyclotron resonance (ECR) plasma source. The Auger lines of Mg KLL and O KLL are converted into atomic densities by cross-calibration using *in situ* XPS intensities of the Mg 2p and O 1s photoelectron lines.

Spectroscopies combined with RHEED 203

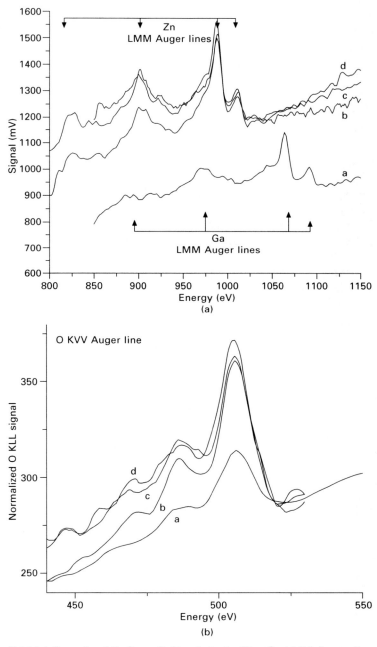

7.11 (a) Growth of ZnO on GaN substrate. The Ga LMM Auger line disappears as the Zn LMM line grows rapidly to a steady state value. (b) Growth of the oxygen O KLL Auger line under the same conditions. Curves labelled 'a' correspond to the initial substrate, curves labelled 'd' indicate the final state.

The atomic percentages of Mg and O remain constant at 50 ± 2% during the growth. Heteroepitaxy of Cr, Mo alloy films on MgO can have a wide range of composition controlled by the atom fluxes. The composition of the films is measured during growth by monitoring the intensity of the Cr LMM and Mo MNN Auger lines. The calibration is performed by measuring the signal of the pure elements Cr and Mo substrate and a linear interpolation (ratio method) is used to quickly determine the film composition. The results are in good agreement with *in situ* XPS calibrations.

In situ growth control of lanthanide alloys

The Auger probe was used by Calley *et al.* (2011) to monitor the co-deposition of Fe, Dy and Tb on silicon substrates using the Auger spectrometer described by Staib (2011). The Auger spectra were measured at a primary beam energy of 10 kV and a beam current in the range of 3–10 µA. The intensities of the Dy and Tb MNN Auger peaks were much weaker than for the Fe lines because the Auger yield issued from the M shell ionization was distributed into a more complex multiplet structure extending over a wider energy range. The detection sensitivity was increased by using a lower energy resolution ΔE = 20 eV. The MNN lines shown in Fig. 7.12 correspond to pure Tb (100%) and

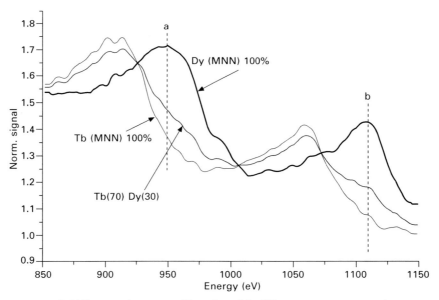

7.12 Dysprosium on a Si wafer with different percent monolayer coverage of terbium ranging from bare Dy to 10% of a monolayer coverage of Tb. The variations of the signal are most pronounced at energies labelled a and b.

pure Dy (100%). The quantification of the signals into atomic concentrations is possible using the peak heights, the area under the peaks, or the derivative signals and measuring the peak to peak amplitudes. These techniques are complicated in this example by the fact that the two Auger lines strongly overlap. The measured energy of an alloy Tb(70) Dy(30) is also shown. It shows that in spite of the overlap, there are energy regions marked by the lines a and b where the distributions vary markedly between the pure element. It is then possible to characterize a specific alloy composition calibrating the signals in this energy range as shown by Calley et al. (2011). A precision better than 2% for the atomic ratios could be demonstrated.

Chemical information from the Auger line shape and energy position

Auger lines provide multiple information about the composition and chemical state of the surface layer. All elements, except H, can be directly identified by their line energy and the concentration of elements can be deduced from the signal strength after calibration. In contrast with X-ray emission lines, the line shape and energy position can provide chemical information about the surface atoms.

The shape of the Auger lines provides useful information about the chemical environment and in-depth distribution of the elements as described by Ramaker (2003). The shape of the peak reflects the details of the transitions between core and valence-type band which depend on the density of states. In addition, the satellite structure of the Auger lines reflects the distribution of CEL that varies with the chemical environment of the atoms. The probability for the occurrence of a CEL depends on the position of the atom with respect to the surface. The line shape of atoms from deeper atomic layers show strong, multiple CEL structures, more visible in the $N(E)$ mode than the $dN(E)/dE$ mode. Auger peak energies are, in many cases, sensitive to the chemical bounding of the surface atoms. Chemical shifts, which are directly related to the changes in binding energy of core electron levels, are more common in XPS. Energy variations of an Auger peak are the result of transitions involving three energy levels which are all shifted by chemical bonding. In addition, the final state is a double ionized atom followed by a relaxation process as described by Grant (2003). The resulting shift is more complex than for XPS and some chemical bonding may not lead to noticeable energy shifts when all involved atomic levels are shifted by the same amount. However, a large number of chemical species result in changes in peak position and shape. The change in the energy of an Auger line was first accurately measured in XPS spectra. The chemical effects are measured using the energy difference between the Auger line and the photoemission line of the same element. The difference is constant if both lines shift with the same amount, but it will change if the Auger line shifts differently. The relation between the chemical shifts of the Auger and XPS peaks was described by Wagner (1972), introducing the notion of Auger parameters. A detailed description of the processes involved was given by Moretti (2003). A useful set of experimental data may be found in older AES reference books, such as from Wagner *et al.* (1979) and Moulder *et al.* (1992). The *in situ* during growth measurement of the Auger has a unique capability to show variations of chemical bonds of elements on the surface. An example of such behavior is given by Staib (2011) where the oxygen Auger peak shows a shoulder that corresponds to adsorbed oxygen on the surface and not incorporated into the oxide matrix.

The measurement of energy shifts can be complicated by the fact that the surface potential may vary as a result of charging up when the surface has a poor electrical conductivity. The energy position of the peak can be shifted by large amounts and, in a worst case, the signal becomes very unstable, making the measurement impossible. For moderate charging effects, the peaks will all be shifted by the same amount. The energy difference between peaks can then be used to measure the relative energy shifts between different Auger lines.

Control of the surface purity during the deposition process

Control of the surface purity during the deposition process is the most common application of AES, an obvious application for monitoring the atomic purity of the deposited layer. The surface composition can be monitored during the growth process. Although the vacuum chamber and deposition sources are most carefully prepared, experience shows that unexpected materials can be deposited on the surface. Figure 7.13 shows the sudden appearance of a phosphorus line when starting a deposition source. The P impurity is soon buried during growth, but the electrical properties of the deposited layer are likely strongly modified. Similar observations have been made on carbon, oxygen and chlorine. The systematic use of AES during growth is a convenient way to guarantee the chemical integrity of the material. This monitoring may become an important tool in performing purity control in real time throughout the deposition process.

7.5 Conclusion and future trends

The different techniques available conjointly with RHEED have been presented. Fortunately, the RHEED electron beam with a grazing incidence angle enables all techniques to achieve higher emission yields and better surface sensitivity. The requirements for the incident beam are very similar for all diagnostics. A beam current adjustable in the range 100 nA to 10 μA is sufficient for almost all applications. The energy range can be limited to

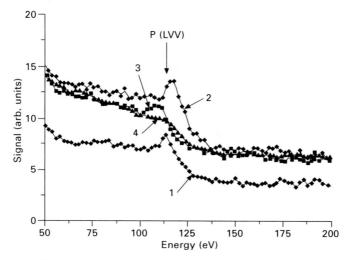

7.13 Phosphorus impurity Auger line observed during Ti oxide growth. The signal growth suddenly deceases as the P atoms become buried under the deposited titanium oxide layer. The impurity appears almost instantly (curve 1), grows further (curve 2) and slowly vanishes (curves 3 and 4).

5–30 keV for most applications, but lower energies in the range 1–5 keV are very useful to acquire REELS data with an AES spectrometer, and for maximizing the emission yield of low-energy X-ray lines by drastically reducing the bremsstrahlung background and the penetration depth of the primary electrons.

The different aspects of *in situ* during growth analytical techniques using the RHEED electron beam are summarized in Table 7.1. The main physical parameters are presented as well as a rating of the performance and cost. The figures and values presented in the table should not be taken too literally, but rather as describing the standard capabilities of the technique, thus making an abstraction of exceptional figures that can be reached only under the most favorable experimental conditions.

For some surfaces, forming known stoichiometric compounds with known atomic concentrations, the calibration of the signals can be performed *in situ*. For alloy materials, quantitative calibrations are commonly improved by performing a cross-calibration measuring reference samples in other carefully calibrated equipment. The most usual techniques for calibration are RBS, XPS, and secondary ion mass spectroscopy (SIMS). XPS devices are often mounted in a separate vacuum chamber connected to the growth chamber that allows *in situ* analyses. Normally, only a few samples need to be cross-calibrated in order to build a set of sensitivity factors.

In spite of their strong potential capabilities, it is likely that the use of the

Table 7.1 Comparison of available in situ, real-time spectroscopies using the RHEED electron beam as excitation. Some values indicated are best estimates based on the author's available information

	XRF or XES	AES and CEL	REELS	PL	TRAX
Spectroscopy name	X-ray fluorescence or emission	Auger electron Characteristic energy loss	Reflection electron energy loss	Photo luminescence	Total reflection angle X-ray
Elemental analysis	Yes, low Z difficult	Yes, all Z >1	No	No	Yes, low Z difficult
Quantitative analysis	Yes, very good for bulk	Yes, very good for surface	Yes, good for surface and bulk	No	Under development
Sensitivity (% element concentration)	Better 5% Down to 0.1%	Better 1% Down to 0.01%	Better 5%	n.a.	1% Down to 0.01%
Band structure information	No	Limited	Yes	Yes	No
Impurity detection	Limited	Excellent	Limited	No	Limited
Analyzed elements	All but low Z, requires special detectors	All Z (except H)	All Z	Limited to light emitting materials	All but low Z, requires special detectors
Depth information	Up to a few μm	Limited to only 0.1–10 nm	Limited to 10–50 nm	Bulk emission to several μm	Limited to 0.1–50 nm
Direct chemical information	None	Peak position, shape	Loss peaks, energy edges	Band structure	Angular distribution
Estimated acquisition time	10s to 10min	10s to 10min	10s to 5min	1–20min	10s to 60 min
Use in UHV vacuum range	Requires detector DP Not bakeable	UHV compatible Bakeable	UHV compatible, some UHV compatible devices need DP	Bakeable	Requires detector DP Not bakeable
Investment costs	$$, $$$ with low Z detector option	$$	$$$	$	$$$, more $ with low Z detector option
Equipment	X-ray detector + acquisition + detector DP	Auger spectrometer + acquisition	High energy resolution filter + acquisition + may require DP	Optical spectrometer + acquisition	X-ray detector + mechanical positioning + acquisition + detector DP

DP = differential pumping; UHV = ultra-high vacuum

above described *in situ* spectroscopies will be limited to specific applications where the growth must not only be controlled by the material fluxes and sample temperature, but also requires control over parameters such as incorporation rate and stoichiometry, knowledge of the layer and interface composition in multilayered samples, or simply control of impurities eventually co-deposited during the growth. For these challenging cases, the above techniques, especially with regards to their new improvements, should become a welcome addition to the usual growth control techniques described in this book.

7.6 Sources of further information and advice

Drouin D, CASINO V2.42 – Program available for free from: http://www.gel.usherbrooke.ca/casino/What.html.

Jablonski A, Tougaard S (2010), 'NIST Elastic Electron Scattering Cross Sections Database', NIST Reference Database 64, US Department of Commerce and Technology.

Mucal, Fluorescence yields and X-ray energies. Available from: http://www.csrri.iit.edu/mucal.html.

NIST database 64, 'Electron Elastic Scattering Cross Sections', Available from: http://physics.nist.gov/asd3 [2010, September 6]. National Institute of Standards and Technology, Gaithersburg, MD.

Ralchenko Y, Kramida AE, Reader J, NIST ASD Team (2008), 'NIST Atomic Spectra Database (version 3.1.5)'. Available from: http://physics.nist.gov/asd3 [2010, September 6]. National Institute of Standards and Technology, Gaithersburg, MD.

7.7 References

Briggs D, Grant J (2003), *Surface Analysis by Auger and X-Ray Photoelectron Spectroscopy*, IM Publications and Surface Spectra Limited, 259.

Calley L, Staib P, Lowder J, Doolittle A (2011) 'An Auger electron analyzer system for *in situ* MBE growth monitoring', *Proceeding European MRS* TuP33.

Chambers SA, Tran TT, Hilman TA (1996), 'Auger electron spectroscopy as a real-time compositional probe in molecular beam epitaxy', *J Vac Sci Tech A*, **13**(1), 83.

Chandril S, Keenan C, Myers TH, Lederman D (2009), '*In situ* thin film and multilayer structural characterization using X-ray fluorescence induced by reflection high energy electron diffraction', *J Appl Phys*, **106**, 024308.

Ding ZJ, Salma K, Li HM, Zhang ZM, Tokesi K, Varga D, Toth J, Goto K, Shimizu R (2006), 'Monte Carlo simulation study of electron interaction with solids and surfaces', *Surf Interface Analysis*, **38**, 657.

Disko MM, Ahn CC, Fultz B (1992), *Transmission Electron Energy Loss Spectrometry in Materials Science*, The Minerals, Metals & Materials Society (TMS).

Drouin D, Real A, Couture D, Joly X, Tastet V, Aimez, Gauvin R (2007), 'CASINO V2.42 – a fast and easy-to-use modeling tool for scanning electron microscopy and microanalysis users', *Scanning*, **29**, 92–101.

Egerton J (1992), *Transmission Electron Energy Loss Spectrometry in Material Science*, The Minerals, Metals & Materials Society (TMS).

Everhart TE (1960), 'Simple theory concerning the reflection of electrons from solids', *J Appl Phys*, **31**, 1483.

Everhart TE, Hoff PH (1971), 'Determination of kilovolt electron energy dissipation vs penetration distance in solid materials', *J Appl Phys*, **42**, 5837.

Ferguson I (1989), *Auger Microprobe Analysis*, Adam Hilger IOP Publishing, 25.

Grant J (2003), in *Surface Analysis by Auger and X-ray Photoelectron Spectroscopy*, Briggs D, Grant J, eds, IM Publications and Surface Spectra Ltd, Chapter 3.

Gryzinski M (1965), 'Two-particle collisions. I. General relations for collisions in the laboratory system', *Phys Rev*, **138**, A305, A322, A336.

Hasegawa S, Ino S, Yamamoto Y, Daimon H (1985), 'Chemical analysis of surfaces by total-reflection-angle X-ray spectroscopy in RHEED experiments (RHEED-TRAXS)', *Jpn J Appl Phys*, **24**, 6, L387.

Hashimoto M, Arkun FE, Jackson A, Clark A, Smith R, Sewell R, Palmstrøm CJ (2009), '*In-situ* compositional analysis of rare earth binary and ternary, oxides by energy dispersive X-ray spectroscopy during MBE growth', *Proceeding NAMBE 2009*, VII.2

Ichimura S, Shimizu R (1981), 'Backscattering correction factor for quantitative analysis', *Surf Sci* **112**, 386.

Ino S, Ichikawa T, Okada S (1980), 'Chemical analysis of surface by fluorescent X-ray spectroscopy using RHEED-SSD method', *Jpn J Appl Phys*, **19**, 1451.

Kanter H (1970), 'Electron mean free path near 2 keV in aluminum', *Phys Rev*, **B1**, 2357.

Krause M O (1979), 'Atomic radiative and radiationless yields for K and L-shells', *J Phys Chem, Ref Data*, **8**, 307 (1079).

Leapman R (1992), 'EELS quantitative analysis', in *Transmission Electron Energy Loss Spectrometry in Material Science*, Disko M M, Ahn C C, Fultz B, eds, Minerals, Metals & Materials Society, Chapter 3.

Lee K, Myers T H (2007), 'The use of cathodoluminescence during molecular beam epitaxy growth of gallium nitride to determine substrate temperature', *J Electronic Mater*, **36**(4), 431.

Lightowlers E C (1990), in *Photoluminescence Characterization, Growth and Characterization of Semiconductors*, RA Stradling, PC Klipstein, eds, Adam Hilger Publisher.

Moretti G (2003), *Surface Analysis by Auger and X-Ray Photoelectron Spectroscopy*, IM Publications and Surface Spectra Limited, 501.

Moulder JF, Stickle WF, Sobol PE, Bomben KD (1992), Appendix A, in *Handbook of X-Ray Photoelectron Spectroscopy*, Perkin-Elmer Corporation.

Murata K (1974), 'Spatial distribution of backscattered electrons in the scanning electron microscope and electron microprobe', *J Appl Phys*, **45**, 4110.

Nikzad S, Ahn CC, Atwater J (1992), 'Quantitative analysis of semiconductor alloy composition during growth by reflection-electron energy loss spectroscopy', *J Vac Sci Tech*, **B10**, 762.

Pellegrino J G, Armstron J, Lowney J, DiCamillo B, Woicik J C (1998), 'Electron beam induced X-ray emission: an *in situ* probe for composition determination during molecular beam epitaxy growth', *Appl Phys Lett*, **73**, 3580.

Powell C (2000), NIST Standard Reference Database 71, 'Electron Inelastic-Mean-Free-Path: Version 1'. Available from: http://www.nist.gov/data/nist71.htm.

Raether H (1980), *Excitation of Plasmons and Interband Transitions by Electrons*, Springer Tracts in Modern Physics, Springer Verlag, Vol 88.

Ramaker DE (2003), *Surface Analysis by Auger and X-Ray Photoelectron Spectroscopy*, IM Publications and Surface Spectra Limited, 465.

Reimer L (1985), *Scanning Electron Microscopy*, Springer Verlag.

Sevier KD (1972), *Low Energy Electron Spectroscopy*, Wiley Interscience, New York, 356.

Sewell PB, Cohen M (1967), 'Reflection high energy electron diffraction and X-ray emission analysis of surfaces and their reaction products', *Appl Phys Lett*, **11**(9), 298.

Small JA, Leigh SD, Newbury DE, Myklebust RL (1987), 'Modeling of the bremsstrahlung radiation produced in pure-element targets by 10–40 keV electrons', *J Appl Phys*, **61**(2), 459.

Sogard MR (1980), 'Backscattered electron energy spectra for thin films from an extension of the Everhart theory', *J Appl Phys*, **51**, 4412.

Staib P. (2011), '*In situ* real time Auger analyses during oxides and alloy growth using a new spectrometer design', *J Vac Sci Technol B*, **29**, 03C125.

Staib P, Tappe W, Contour JP (1999), 'Imaging energy analyzer for RHEED: energy filtered diffraction patterns and *in situ* electron energy loss spectroscopy', *J Cryst Growth*, **201/202**, 45–49.

Strawbridge B, Shinh RK, Beach C, Mahajan S, Newman N (2006), 'Effect of surface topography on reflection electron energy loss plasmon spectra of group III metals', *J Vac Sci Technol A*, **24**(5), 1776.

Sun B, Goodrich TL, Ziemer KS (2009), 'Using RHEED-TRAXS to understand complex oxide growth mechanisms', *Proceeding NAMBE 2009*.

Tanuma S (2003), in *Surface Analysis by Auger and X-Ray Photoelectron Spectroscopy*, D Briggs and J Grant, eds, IM Publications and Surface Spectra Limited, 259.

Tompkins RP, VanMil BL, Schires ED, Lee K, Chye Y, Lederman D, Myers TH (2006), '*In-situ* investigation of surface stoichiometry during InGaN and GaN growth by plasma-assisted molecular beam epitaxy using RHEED-TRAXS', *Mat Res Soc Symp Proc*, **892**, 0892-FF04-06.1–5.

Wagner CD (1972), 'Auger lines in X-ray photoelectron spectrometry', *Anal Chem*, **44**(6), 967.

Wagner CD, Riggs WM, Davis LE, Moulder JF, Mullenberg GE (1979), *Handbook of X-Ray Photoelectron Spectroscopy*, 1st edn, Perkin Elmer Corporation (Physical Electronics).

Wang ZL (1996), *Reflection Electron Microscopy and Spectroscopy for Surface Analysis*, Cambridge University Press.

Werner S (2003), in *Surface Analysis by Auger and X-ray Photoelectron Spectroscopy*, D Briggs and J Grant, eds, IM Publications and Surface Spectra Limited, Chapter 10.

Werner S (2010), 'Electron Transport in Solids for Quantitative Surface Analysis'. Available from: http://eaps4.iap.tuwien.ac.at/~werner/Si_pl.dat.

Williams G (2006), 'Electron Energy Level Binding Energies'. Available from: www.dovada.com/electron_binding.htm.

ZAF from NIST. Available from: http://www.cstl.nist.gov/div837/Division/outputs/DTSA/chapters/Analysis.html#DZ Lecture on ZAF, Available from: http://web.pdx.edu/~jiaoj/phy451/Lect8.pdf.

8
In situ deposition vapor monitoring

V. MATIAS, Los Alamos National Laboratory, USA and
R. H. HAMMOND, Stanford University, USA

Abstract: *In situ* deposition vapor monitoring is essential for accurate control of film thickness and elemental composition during physical vapor deposition. This chapter describes current state-of-the-art vapor sensing technologies and compares their relative merits. We review the quartz crystal microbalance, vapor ionization and optical absorption measurement techniques.

Key words: deposition rate monitoring, vapor flux monitor, quartz crystal microbalance, electron-impact emission spectroscopy, quadrupole mass spectroscopy, atomic absorption spectroscopy.

8.1 Introduction

State-of-the-art thin-film deposition seeks to achieve atomic-level control of thickness and tuning of atomic composition to better than 1% during co-deposition of several species. These demands require very precisely calibrated atomic fluxes that are also controlled in real time using vapor flux sensors. Deposition vapor sensors and control systems have evolved over the last several decades and several different types are now available commercially that satisfy these requirements. Still more deposition monitors pushing the limits in rate control are in the realm of ongoing research.

This chapter is intended primarily as a review of advanced vapor flux monitors for practitioners of thin film deposition. We review the technologies that are commonly used for vapor flux sensing in physical vapor deposition (PVD). We discuss each technique and compare their relative merits, their sampling times and sensitivity, as well as equipment and methods used, and future trends. The reader is referred to further literature on each topic and to sources of commercially available deposition rate monitors. We also discuss techniques, such as tunable-diode laser atomic absorption spectroscopy, that have been developed, but are not as yet available commercially.

8.2 Overview of vapor flux monitoring

Vapor flux monitoring can be accomplished with a variety of vapor sensors. The review article by Buzea and Robbie (2005) describes some of these techniques. Accurate monitoring and control of deposition rates, however, necessitates from the start a proper design of the deposition system to include the requirements of the desired sensors. Certain sensors need to be accommodated with adequate space within the deposition chamber and others require precisely positioned ports for optical access. Figure 8.1 shows a generic PVD chamber schematic with some possible sensors. Inclusion of most sensors in the chamber is much simpler if it is conceived during the early design stage and before the system is built.

We focus on measuring the deposition rate from the vapor flux. The atomic flux, or incidence rate, in a vapor (number of atoms per unit area per unit time) in a certain direction is commonly defined as:

$$\Gamma_{in} = N \langle v_z \rangle$$

where N is the atomic number density of the deposition vapor and $\langle v_z \rangle$ is the mean velocity of the atoms in that direction. For a true film growth rate R in nm/s (Jaccard et al., 1994) on the substrate one needs to multiply the atomic flux by the atomic mass m and the sticking coefficient (fraction of atoms that remain on the substrate after deposition) k and divide by the density of the film, ρ_f (mass per unit volume):

$$R = k \cdot m \cdot \Gamma_{in} / \rho_f$$

Typically with a deposition sensor one measures an atomic vapor density or a mass flux, but the sticking coefficient on the substrate needs to be deduced separately by measuring the thickness of the deposited film on the substrate. This chapter discusses only *in situ* measurements of the deposition vapor and not the film thickness measurement techniques. RHEED, for example, discussed in Chapters 1 and 7, and optical reflection intensity or phase

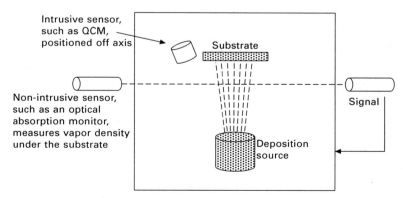

8.1 Schematic of a PVD vacuum system with several sensors for vapor flux monitoring. The vapor flux signal from the sensors is fed back to the deposition source for regulation in order to maintain a constant deposition rate (QCM = quartz crystal microbalance).

measurements from the film, discussed in Chapter 5, can be used *in situ* to calibrate the thickness of the deposited film and hence the sticking coefficient. A profilometer and transmission electron microscopy (TEM) cross-sections are among *ex situ* tools that can also be used to deduce film thickness.

Measurement of the deposition rate can be either intrusive, by placing the sensor inside the vapor and obstructing deposition in the line from the source, or non-intrusive by performing a remote sensing measurement such as optical absorption through the vapor; see Fig. 8.1.

A deposition source has a deposition distribution that can typically be modeled as a $\cos^n(\theta)$ dependence where θ is the angle from the source direction (Smith, 1995). The exponent n is often found to be between 1 and 4, depending on the type of source and source condition. Intrusive vapor sensing techniques suffer the disadvantage that the flux is measured at a different position, i.e. angle, than the substrate and hence need a correction factor, commonly called the 'tooling factor', for proper calibration to the vapor flux at the substrate. More significantly, the flux profile can change during deposition time and thus the calibration may need to be done periodically or even continuously. It is of course desirable to measure the vapor flux directly under the substrate, i.e. non-intrusively. Although non-intrusive techniques, such as optical sensing, are preferred, the field has suffered from a lack of commercially available sensors.

We differentiate also between techniques that measure mass fluxes and/or mass accumulation directly and ones that measure the density of the vapor. Typically a quartz crystal microbalance (QCM) is used for a mass flux measurement. Most other vapor sensors measure only the vapor density. The other division of sensors is whether they are atomic species specific or not. QCM is not species specific, but atomic absorption (AA), for example, is.

Techniques that can measure atomic species in the vapor quickly, accurately, and non-invasively are the best ones from a rate control perspective. This includes several atomic absorption techniques, which we discuss in more detail. The bandwidth requirement for control depends on the type of source and rate that is used. Typical effusion cells where temperature is kept constant require flux sampling times that are tens of seconds or longer, and similarly for sputtering sources where the power is kept constant. On the other hand electron-beam evaporators may require fast feedback in the scale of milliseconds. Thus different feedback bandwidths are required depending on the system, but in each case the sensor signal can be fed back in a loop to keep the vapor flux constant.

8.3 Quartz crystal microbalance (QCM)

QCM for mass flux measurement is the most powerful and widely used technique in vapor sensing. QCMs are used in most industrial processes for rate control where rate control is implemented. QCM is also unique among the techniques discussed in this chapter in that it directly measures mass per unit area deposited from a vapor, assuming that one knows the elastic property (acoustic impedance) of the material. Unlike other techniques, it needs no further calibrations. Furthermore if the film density and the sticking coefficient on the sample are known, then the deposition rate can be precisely calculated.

A QCM consists of an oscillating piezoelectric quartz crystal that is sensitive to the added mass, as shown in Fig. 8.2. When a voltage is applied across the sides of a piezo crystal it will resonate at discrete frequencies. Then when mass is added to the face of a resonating crystal, the resonant frequency is reduced. This change in frequency is very repeatable and well understood for specific oscillations of the quartz, such as for the quartz AT cut. The AT-cut quartz, which vibrates in its thickness shear mode, is the most commonly used crystal monitor because of its superior temperature stability and good mass sensitivity. Typically, one surface of the crystal is made in a spherical shape to contain the vibrational energy within the center of the crystal and to reduce the coupling to unwanted modes at the edges (Shockley *et al.*, 1963; Wajid, 1997).

In 1959 Sauerbrey first described the equation for the change in frequency, Δf, of an oscillating quartz crystal after coating with a certain added mass M_f,

$$\frac{M_f}{M_q} = \frac{-\Delta f}{f_q}$$

where M_q is the mass of the uncoated quartz crystal and f_q is its uncoated resonant frequency. The first frequency measurement instruments used the

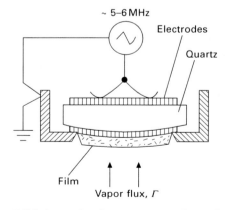

8.2 Schematic of a quartz cystal monitor.

following simple equation, known as the Sauerbrey equation, for the deposited film thickness on a QCM, d_f:

$$d_f = -\frac{N_{AT} \rho_q}{f_q^2} \frac{\Delta f}{\rho_f}$$

where ρ_f and ρ_q are the densities of the film and quartz crystal (2.649 g/cm^3), respectively, and N_{AT} is the frequency constant of an AT cut quartz crystal (166,100 Hz cm).

If the change in resonant frequency is greater than 2% of f_q the above equation is not valid and one needs to take into account the acoustic impedance mismatch between the quartz and the film. The resulting equation that treats the problem as a one-dimensional composite resonator was developed by Lu and Lewis (1972):

$$d_f = -\left(\frac{N_{AT} d_q}{\pi \tau_f f_q Z}\right) \arctan\left[Z \tan\left(\frac{-\pi \Delta f}{f_q}\right)\right]$$

where Z is the acoustic impedance ratio $Z = (d_q u_q / d_f u_f)^{1/2}$ and u_q and u_f are the shear moduli of the quartz and film, respectively. Advances in microprocessors taking place at the same time in the 1970s made it practical to solve the full equation in real time with the instrument. Most QCM deposition process controllers sold today use this sophisticated equation for calculating the film thickness.

Typical instrumentation prior to the mid-1990s relied on the use of an active oscillator circuit that actively keeps the crystal in resonance. Oscillation in the crystal is then sustained as long as the gain provided by the amplifiers is sufficient to offset losses in the crystal and the circuit. As the crystal is loaded with more mass the electrical characteristics of the oscillator circuit degrade, the impedance Z rises, and the system becomes unstable to mode hopping.

A new type of system developed by Wajid (1996), using a so-called 'mode-lock technology,' eliminates the crystal from the active oscillator and instead has a separate resonating circuit that allows the instrument to constantly test the crystal response to an applied frequency. In this way the instrument can determine the resonant frequency and at the same time verify that the crystal is oscillating in the desired mode. The new system is immune to mode hopping and resulting inaccuracies. The crystal's frequency can be determined to less than 0.005 Hz at a rate of 10 times per second. Furthermore since this system can measure several modes as it sweeps the frequency, other features are available. In particular one can measure the frequency shift of the lower frequency anharmonic that is close to the fundamental mode. This mode has a slightly different mass sensitivity from the fundamental mode and this difference can be used to estimate the Z-ratio of the material. This is the so-called automatic Z-ratio instrument calculation performed in the advanced QCM controllers today.

Among the most common applications of QCM technology are in the production of optical coatings. For these dielectric materials the useful crystal life is much shorter than for deposition of metals. Unlike typical metal films where the crystal failure is due to the viscous losses in the bulk of the material, in this case stress in the thin film causes cracking and the rough surface breaks up the oscillatory wave of the quartz crystal which causes loss of resonance. Wajid (1993) reported significant improvements in the life of the quartz crystal by adding a compliant buffer layer between the gold electrode and the crystal. This buffer layer needs to have good adhesion and low acoustic damping and one such material is zinc. Observed crystal life enhancements were two to six-fold.

Still one of the main drawbacks of the QCM is the finite lifetime of the quartz oscillator. The crystal is continuously being coated with a damping material and the resonant Q decreases continuously; eventually the resonant mode cannot be measured. To help alleviate this problem, manufacturers of QCM systems have developed 6-, 12- and even 24-crystal carriers where the crystals in reserve can be automatically switched in during the process. For long continuous deposition processes, as in web coating, however, the finite QCM life can still be a significant limitation.

Another significant disadvantage is that the QCM is not species specific, i.e. it cannot differentiate between different atoms deposited or potential reactions taking place on the QCM such as oxidation of the atoms. On the other hand, when properly calibrated, the QCM is a very powerful tool and can also be used to measure reactions, such as film oxidation on the crystal or etch rates, and consequently can also be used to measure reactive gas fluxes; see for example Matijasevic *et al.* (1990).

8.4 Vapor ionization techniques

Most vapor flux sensing techniques other than the QCM utilize excitation of the vapor molecules to detect their response. Vapor ionization techniques, for example, ionize the atoms in the vapor plume in order to measure the atomic density. The ionization rate and the resulting collected ion current are proportional to the atomic density of the vapor in the sensor, N. Figure 8.3 shows a schematic of a generic device of this type.

8.4.1 Ion gauge and chopped ion gauge (CIG)

The simplest type of an ionization sensor is just an ion gauge. In an ion-gauge flux monitor the filament should be shielded so the vapor does not react with the filament and other components should be shielded from accumulating

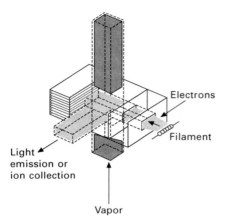

8.3 Schematic of a generic vapor ionization sensor. The ionized atoms can be collected or mass filtered, or light emission can be analyzed.

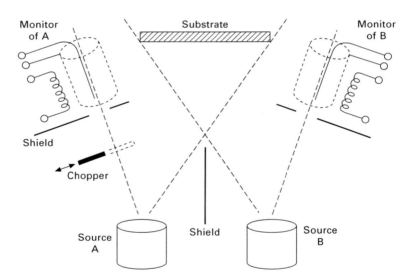

8.4 Schematic of ion gauges used in co-deposition.

deposits. Simultaneous monitoring during co-deposition can be achieved by shielding each monitor from the other sources, as shown in Fig. 8.4.

In order for the ion gauge sensor to be sensitive to the deposition flux only and not to the background gas, a mechanical chopper wheel can be used; see Fig. 8.4. This is the so-called 'chopped ion gauge' (CIG) monitor. The deposition flux that enters the gauge is then pulsed by the chopper and can be measured with a phase-sensitive lock-in detector.

The CIG has seen extensive use in basic research over the past 45 years, particularly for superconducting compounds (A-15 and high-temperature superconducting cuprates) and for industrial optical coating on a large scale (Hammond, 1975, 1978). As large-scale processes requiring electron beam sources for their high rates become attractive, the knowledge and experience of the unique control features needed for high rate are becoming important. Initially after the introduction of the CIG the bearings in the motor and chopper made the unit unreliable. However, the advent of modern high-vacuum high-temperature greases have eliminated this issue.

The lessons learned from the use of the CIG can be valuable for controlling processes involving high-rate multiple source co-evaporation, especially in cases where at least one of the elements is a high melting point metal. This knowledge can be applied to other rate monitors that sample individual evaporation sources, either by collimation or by element specificity. Niobium and similar transition metals are such an example where suitable crucibles or liners are not available. The evaporation of such materials from a water-cooled bare hearth by the electron beam heating of the surface results in a rate, as measured at one point in space, having random fluctuations in time. The amplitude of the fluctuations may be 50% of the rate, with frequency components up to tens of Hz. The source of this variation is easily seen upon looking at the molten source material using an optical filter (welding glass): the large gradient in the temperature between the impact point of the electron beam and the region in contact with the water-cooled copper hearth produces violent convection turbulence in the molten material. The brightest regions, which correspond to the regions with the highest rates of evaporation, are seen to move around and change shape.

Co-condensation of atoms from two sources would, under the conditions just described, produce an inhomogeneous composition. The CIG evaporation rate monitors that were developed to solve the above problem are based on the ionization gauge, with features to permit its use with electron beam evaporation. The sensor has the Bayard–Alpert configuration, i.e., the filament outside of the helical grid, and the collector inside. This permits the majority of the evaporant to pass through the gauge without condensing onto the elements of the gauge (the collector is offset from the center of the grid axis, as is the collimated evaporant flux). Charged particles of both signs are present in electron beam systems, and can cause erroneous signals on the collector. Electric and magnetic stripping are used to mitigate this issue. Small bar or horseshoe magnets usually are sufficient to deflect the electrons that are present at the energy corresponding to the high voltage at the gun. The secondary electrons produced are captured by the electric strippers with ±30V. In order that the sensor be sensitive to the evaporant flux only, and not to the background gas also, a mechanical wheel chopper is used. The evaporant flux that enters the gauge is thus pulsed and phase locked to the chopper. The AC signal, which is proportional to the rate, can be measured with a phase-sensitive lock-in detector. This is compared with a pre-set DC level, which is set proportional to the desired steady rate of evaporation. The difference signal, an error signal of positive or negative voltage, is fed back to the electron beam power supply.

The error signal is used in two ways at the electron beam power supply to provide a feedback to the evaporation rate. The first and most commonly used method is to use the error signal to vary the electron gun emission by changing the temperature of the filament by controlling the emitter current. This is satisfactory for error signals whose main frequency components are less than 1 to ½ Hz. The electron gun filament temperature response begins to get out of phase with the change in the heating current at about 1 Hz. This results in the feedback changing from negative to positive, and wild oscillations result.

A second feedback loop has been devised to deal with the higher frequency fluctuations. Lateral deflection magnets (as part of the e-gun source) are used as follows: a triangular, or other, current wave-form at about 50 Hz is used to move the electron beam from side-to-side at the point of impact on the molten metal being evaporated. The effect of moving the beam is to decrease the average power density, thereby reducing the evaporation rate. The amplitude of the beam motion is adjusted such as to reduce the rate by 5 to 20%. The error signal from the CIG is then used to modulate the amplitude of the sweep, thus changing the evaporation rate. The advantage of this method is that it is capable of responding to changes in the rate at frequencies in the 1–10 Hz range. A similar method can be used by controlling the sweep amplitude on the Wehnelt grid in the electron emitter of sources so equipped. In that case one can extend the feedback bandwidth to 100s of Hz. The long-term stability of the CIG-controlled evaporation rate can be better than 2%.

8.4.2 Quadrupole mass spectroscopy

In addition to the ion-gauge monitors described, it is even better, especially in co-deposition, to use a species-specific monitor. One such device is a quadrupole mass spectrometer (QMS), which is simply an ion gauge followed by a mass/charge-selective filter. The QMS has the advantage of high sensitivity and large bandwidth, down to 0.1 pm/s at 1 Hz. Typically the QMS is equipped with an electrometer as a current multiplier and its bandwidth can be 100s of Hz. The key issue with a QMS as a rate sensor is the long-term stability.

Schellingerhout *et al.* (1989) described a control system using a Balzers QMS for e-beam evaporators. They passed the control signal through a low-pass filter to control the filament current and a high-pass filter to obtain a control for the fast Wehnelt grid voltage. Jaccard *et al.* (1994) used a Hiden QMS to precisely calibrate several monolayer deposition of each species in a sequential deposition molecular beam epitaxy (MBE) process. Newer versions of the QMS allow for multiplexing of up to 16 outputs for different masses in feedback control.

The QMS method is becoming especially useful for molecular vapors that need to be cracked to the monoatomic state in order to prepare films of high-quality materials, such as, for example, the topological insulators (such as Bi_2Se_3), and in commercial production of the solar photovoltaic CIGS (CuInGaSe) and other materials containing the molecular elements (As, Sb, S, Se, and Te). These materials as molecules (As_4, Se_8, etc.) in their normal evaporative state do not stick readily or combine in compounds, and require orders of magnitude larger flux than if the atomic species were supplied. Insufficient and inconsistent molecular cracking results in non-stoichiometric compounds, difficulty in achieving the best combination of flux and substrate growth temperature, as well as an enormous consumption of material increasing the cost of production, as well loading the chamber. Methods of cracking the molecules are either thermal or RF plasma. The thermal cracking often results in the dimer (As_2, Se_2, etc.) and does achieve better quality and utilization. On the other hand RF plasma cracking results in the monoatomic state and has much superior sticking capacity and reactivity to form high-quality compounds.

For the consideration of the usefulness of the QMS for monitoring these cracked molecules one needs to understand what the QMS does to the molecules: the ionizer uses electron impact to ionize the elements, typically at 70 V, because this is close to the maximum in the ionization cross-section, in general across the periodic table. In the case of molecules there are additional pathways of producing an ionized particle that is extracted and displayed as the charge/mass ratio in the operation of the spectrometer. These include a two-step process, disassociation followed by ionization, and also a one-step ionization during disassociation. Since the latter involves only one step it appears to have the larger probability. This becomes important because the energy dependence of the cross-section as the energy is lowered close to the threshold (typically 10 eV) allows the ionization during disassociation to be removed at roughly 15 V in the case of As, leaving only the molecule visible in the spectrum without the false impression that cracking has produced atomic species. Thus As without cracking and an electron energy of 70 V will show disassociation-ionization for all molecular sub-units from 4 down to 1. At the lower ionization energy of 15 V only the As_4 ion is found. With thermal cracking As_2 and As_1 appear at 70 V, while at 15 V only the As_2 is seen (Campion *et al.*, 2010). This allows for a good evaluation of the cracking efficiency. The same possibility is indicated by the cross-sectional data for the molecules O_2, N_2, H_2, and carbon oxide (Kieffer and Dunn, 1966).

It is expected that in the case of radio frequency (RF) plasma cracking of the other molecules (in addition to As) the same methods will show the efficiency of the cracking, using the lowered electron energy (soft ionization), and can be used in monitoring and controlling the rates. There are unpublished data indicating that this is the case.

8.4.3 Electron impact emission spectrometry (EIES)

Another species-specific ionization monitor is the electron impact emission spectrometer (EIES). EIES uses the characteristic atomic emission to identify the element being monitored. As in a QMS, an electron beam, in this case of 180 V, is oriented perpendicular to the flux, and upon impact with the atoms excites them to higher levels, which then decay to lower or ground state, emitting photons of characteristic wavelengths (see Fig. 8.3). The intensity of the particular emission summed over all the atoms is linearly proportional to the vapor density of the corresponding species. The following equation relates the light intensity measured for a specific transition I_{ij} to the vapor number density N:

$$I_{ij} = KN\sigma_{ij}I/e$$

where K is a calibration constant relating all the optical system losses, σ_{ij} is the transition cross-section at constant excitation energy, I is the electron current density, and e is the electronic charge.

The light emitted is guided through a vacuum window to a filter or spectrometer, and then the intensity is measured by a photomultiplier tube (PMT) (Lu et al., 1977; Gogol and Reagan, 1983). To avoid interference when evaporating a number of elements, the light path can be divided using split optical fiber bundles and individually chosen optical filters are inserted before independent PMTs. The gain for each PMT can be controlled to distinguish the elements and control the power to the evaporation sources. In practice as many as four channels have been used.

There can be interference among the spectra of several elements when co-depositing elements. This mainly depends on the specifications of the wavelength selection device (filter or monochromator) and the elements of interest. However, in most of the co-deposition cases, one can select appropriate lines to avoid interference, although sometimes this prevents the use of the strongest emission line for certain elements.

The broadband emission from residual gases may also be a problem in certain cases, particularly for low rate deposition in a relatively high background pressure (>10^{-5} torr). However, for non-reactive MBE processes, the residual gas interference is not a serious problem, nor for some industrial processes running at a very high rate with non-ultra-high vacuum (UHV) conditions. In those cases where gas interference does become a problem, particularly in reactive depositions, a dual-chamber EIES sensor can be used to solve the problem (Lu et al., 2008). One chamber has the vapor flux passing through, while the other is open only to background gases. The emitted spectra are added and using appropriate signal processing technique, the interfering signals from the residual gases can be completely eliminated from the output signal.

All the above electron-beam ionization monitoring techniques can be used only when the electron mean free path is larger than the monitor diameter, i.e., in high vacuum with a pressure of less than 10^{-4} torr. The sensors can be differentially pumped and thus could support an order of magnitude higher pressure in the chamber. However, scattering in the chamber at pressures of mid-10^{-5} torr range will change the sensitivity for these instruments and render them useless (Appelboom et al., 1993). The sensors are also not usable in a plasma environment because of interference. For higher pressure processes or in plasmas, optical excitation and detection techniques must be used to monitor vapor concentration.

The ionization monitors typically have a lifetime that is equivalent to hundreds of QCMs. They still need to be cleaned out periodically, but they are effective when depositing large amounts of material continuously. QMS and EIES are species specific as well.

8.5 Optical absorption spectroscopy techniques

Use of optical methods for measuring AA in the vapor was first demonstrated for barium (553.6 nm) by Gunter Wessel (1949). It was as early as the late 1960s that Stirling and Westwood (1969) and Fazekas et al. (1972) demonstrated the use of AA in vapor deposition processes. The technique was thus demonstrated more than 40 years ago, but instrumentation development had a long and complicated path.

Theoretically the fraction of an optical beam remaining after propagating through a medium with absorption coefficient α is given by Beer's law:

$$I(z) = I(0)\exp(-\alpha z)$$

where α has units of 1/cm and is a function of light frequency. The absorption coefficient is proportional to the density of absorbing species N (1/cm^3) and can be expressed as $\alpha = N\sigma$, and the effective absorption path length per pass of length l is $N\sigma l$. σ is the absorption cross-section in units of area (cm^2).

Infrared radiation usually probes molecular vibrations, while visible/UV light probes transitions between different electronic states. Electronic transitions can be to predissociative or dissociative states, leading to a mild or extensive broadening of the absorption lineshape. Moreover, excitation to a level that dissociates strongly perturbs the deposition process. Typical UV absorption cross-sections are $\sigma \sim 10^{-19}$–10^{-16} cm^2. For $\sigma = 10^{-17}$ cm^2, this means that the absorbed fraction ($N\sigma l$) is 1% for a density of $10^{14}/$ cm^3 over a 10 cm path length. Molecules with strong bands at >200 nm can be detected by using standard UV quartz optics, while those that are only <200 nm require vacuum UV techniques and are usually not practical, except in special cases to be discussed later.

There are a number of different light sources that can be used to measure absorption. For example, a tungsten lamp in the optical and a deuterium lamp in the UV can be used, or alternately a Xe arc lamp. Of particular interest are atomic resonance lamps known as 'hollow cathode lamps' (HCL) that are available for most elements. They emit resonance light in the visible and near-UV that can excite their ground state atoms in the vapor. In the next section we discuss the sensors based on these lamps. Another possible light source is the tunable diode laser (TDL). These lasers are able to tune through a narrow range of wavelengths (typically 2–10 nm) and if there is a fortuitous match with the atomic resonance of the species of interest then a sensor can be obtained. Some atomic gas species such as halogens, O, and N absorb in the vacuum ultraviolet (VUV) (110–160 nm) which requires vacuum enclosures of the optical paths.

Figure 8.5 shows schematically what happens during the vapor absorption process for the two light sources discussed here. The HCL produces a broad lineshape caused by Doppler broadening from the lamp emission profile (~2 GHz) compared to the evaporated atoms (~0.5 GHz). The HCL-AA systems give a number that is equal to the integral of the signal lineshape

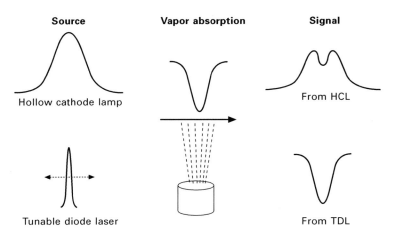

8.5 Comparison of the hollow cathode lamp (HCL) source and tunable-diode laser (TDL) source AA. With an HCL-AA system the source is broader than the absorption line and one only gets an absorption value that does not reflect pure Beer's law. With TDL-AAS the source is much narrower and one can obtain a complete absorption spectrum that can be fitted to Beer's law.

after absorption. Absolute calibration of the atomic density is complicated by deviations from Beer's law in HCL-AA systems. TDL light sources, on the other hand, produce an extremely narrow lineshape (0.3 MHz) that can be swept across the vapor absorption line; typical tuning range is 100 GHz. We call this technique TDL-atomic absorption spectroscopy (TDL-AAS), emphasizing that it is spectroscopic in nature, unlike HCL-AA which only measures an absorption integral. Figure 8.5 shows a comparison of the TDL with HCL sources. The TDL-AAS signal can be fitted to Beer's law exactly as long as the light intensity is not too high causing absorption saturation, also known as bleaching.

8.5.1 HCL-based AA vapor monitors

AA vapor sensing using HCL has been used in several types of deposition processes including sputtering, plasma-assisted deposition, laser-assisted deposition, and more common evaporation. The first AA commercial instruments were developed by ULVAC (Japan) starting in the late 1970s. Following that, Lu *et al.* demonstrated an HCL-AA instrument for deposition rate monitoring in the late 1980s (Lu *et al.*, 1989; Missert *et al.*, 1989). They used AA for evaporation rate control during co-deposition of Y, Ba, and Cu for *in situ* growth of the high temperature superconducting films. Figure 8.6 shows a schematic of the MBE system that was used with a simple

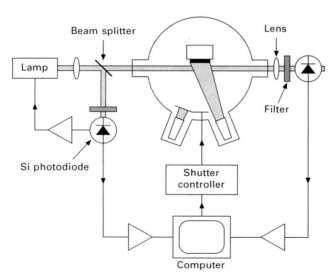

8.6 Schematic of the apparatus to monitor flux by absorption and consequently MBE growth rate. Reprinted with permission from Chalmers and Killeen, *Applied Physics Letters*, **63**, 3131–3133 (1993). Copyright 1993, American Institute of Physics.

AA vapor monitoring by Chalmers and Killeen somewhat later (Chalmers and Killeen, 1993). The authors used the optical-based flux monitoring to control the MBE source shutter operation by timing the shutter opening for the desired amount of deposited material.

Klausmeier-Brown et al. (1992) used a similar system and measured the atomic flux of Bi, Sr, Ca, and Cu by AA during atomic layer-by-layer MBE growth of high-temperature superconducting films. Benerofe et al. (1994) developed a dual-beam atomic absorption system to measure Sr and Ru vapor during co-evaporation of $SrRuO_3$. The reference arm path compensates for light fluctuations from the lamp, and the normalized absorption signal could be fed back to the source for stable control of evaporant flux, as shown in Fig. 8.7. These methods are stable and fast (~200 ms), but sensitivity depends on the element.

Lu and Guan (1995) have described a different dual beam scheme, as shown in Fig. 8.8, that improves the baseline stability by correcting for viewport coating and other optical alignment shifts with a reference xenon lamp (filtered by a monochromator) through the same reactor and reference paths as the hollow cathode lamp. The two lamps are individually controlled by a timing circuit.

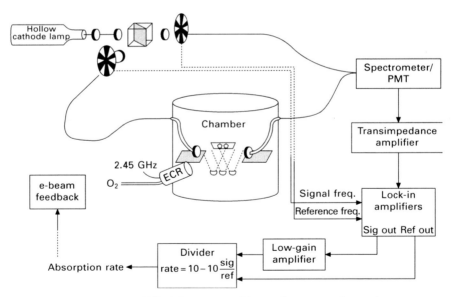

8.7 Dual-beam HCL-AA system with a reference path used to correct the lamp fluctuations (ECR = electron cyclotron resonance). Reprinted with permission from Benerofe et al., *Journal of Vacuum Science and Technology – Section B*, **12** (2) pp. 1217–1220 (1994). Copyright 1994, American Vacuum Society.

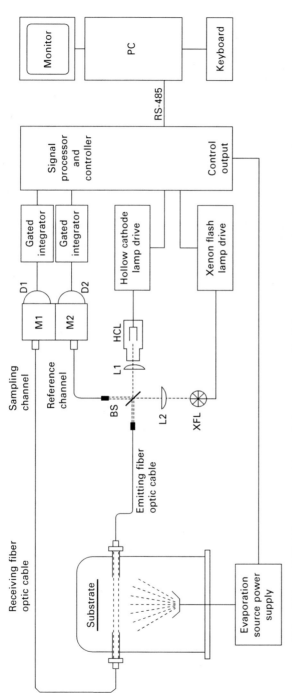

8.8 Dual beam HCL-AA system with a reference xenon flash lamp to correct for viewport coating and optical alignment shifts. The system uses a common optical path for automatic correction of transmission. In addition to the HCL, a xenon flash lamp (XFL) is used for determining optical transmission. Light from both sources is combined in a beam splitter (BS) and then focused onto the fiber optic cable by quartz lenses (L1 and L2). The fiber optic cable guides the light into the chamber and on the opposite side a receiving probe collects the light to a second fiber optic cable. A monochromator (M1) is used to isolate the specific atomic emission line. A silicon photodetector (D1) is used to measure the light intensity. A similar photodetection arrangement (M2 and D2) is used to analyze a portion of the light from each source coming off of the beam splitter and determines the reference channel. All the signal intensities are compared in a microprocessor-based signal processing unit. Reprinted with permission from Lu and Guan, *Journal of Vacuum Science and Technology – Section A*, **13** (2) pp. 1797 (1995). Copyright 1995, American Vacuum Society.

8.5.2 TDL absorption spectroscopy

As already described, tunable diode lasers offer very narrow coherent light sources that can be swept across the absorption line of the vapor to obtain a complete absorption spectrum. This technique was demonstrated extensively in the 1990s, in particular in a collaboration between groups at Stanford University and New Focus, Inc. (Wang et al., 1995b, 1996, 1997). They demonstrated frequency modulation, frequency doubling of the laser wavelength using second harmonic generation in a waveguide, and atomic velocity measurements using two light beams.

Sensitivity in measuring small absorbed fractions can be improved dramatically by frequency modulation (FM) of the laser frequency and homodyne detection of the signal photocurrent at the modulation frequency ω_m or a harmonic. This technique has been employed for TDL-AAS by Wang et al. (1995a), as shown in Fig. 8.9. Laser source noise becomes very small with detection at these high modulation frequencies. Frequency modulation can be achieved by adding an FM component to the DC injection current in the diode laser. When ω_m is less than the frequency half-width of the absorption line, this method is commonly called wavelength-modulation (WM) spectroscopy or derivative spectroscopy. In that case the modulation is usually applied by use of an external electro-optic modulator.

Since the laser lineshape is very narrow, it can exactly reproduce the AA

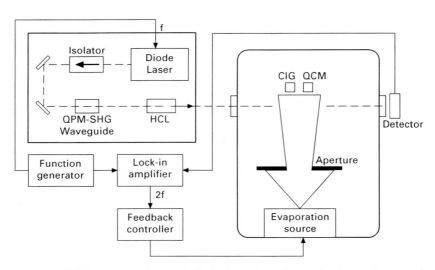

8.9 Frequency doubled-diode-laser-based atomic absorption monitor (QPM-SHG = quasi-phase matched second harmonic generation device). Reprinted with permission from Wang et al., *Applied Physics Letters*, **68** pp. 729–731 (1996). Copyright 1996, American Institute of Physics.

in the vapor plume and an accurate spectrum can be obtained. By measuring the absorption along the direction of the expanding vapor, one can obtain a component of the directional velocity and measure the corresponding Doppler shift. The real atomic plume has a finite angular spread in space, and the velocity has a distribution, resulting in a spectral spread in the absorption profile due to the Doppler broadening. By measuring two spectra, one with a light beam with a component in the direction of the velocity and one against it, two peaks can be measured whose separation will be the average velocity times the cosine of the angle at which the light intersects the beam. Figure 8.10 shows two different schemes: cross-beam and counter-propagating-beam scheme (Wang et al., 1999). The cross-beam scheme is better in principle as separate spectra that do not need to be fitted are obtained. The counter-propagating beam scheme is simpler to implement in a chamber and is usually sufficient if the peaks are well separated. Both schemes have been demonstrated to work in implementing flux measurements (Matijasevic and Slycke, 1998; Wang et al., 1999).

AA vapor monitoring techniques have no limitation on process operating pressure, are non-intrusive, i.e., they can measure right in front of the sample surface, and are atomic species specific. With proper vacuum system design the sampling region can be extremely close to the sample surface for precise control of film composition and thickness, something that is not possible with other techniques. These are strong advantages of AA techniques. However, the effectiveness of a particular AA monitor still depends on the details of the specific atomic transition (absorption coefficient) that is used and on the application.

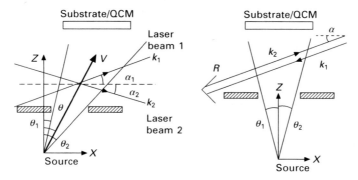

8.10 Two different schemes for measuring the Doppler shift with two laser beams traversing the vapor at an angle. Reprinted with permission from Wang et al., *Journal of Vacuum Science and Technology - Section A*, **17** (5) p. 2676 (1999). Copyright 1999, American Vacuum Society.

8.6 Summary of techniques and resources

Table 8.1 summarizes the characteristics of the six sensors discussed in detail in this chapter. Note that only the QCM measures a true mass flux. Furthermore, only the optical absorption techniques are non-intrusive.

The following is a partial list of commercial suppliers of deposition rate instruments:

- QCM controllers: Inficon (www.inficon.com), East Syracuse, New York, supplies a whole range of QCM controllers including IC/5 and IC6 instruments that are capable of doing auto-Z correction and accurate measurement of fluxes down to 0.01 Å/s.; Maxtech and Sigma Instruments QCM controllers now sold through Inficon; Sycon, (www.sycon.com), East Syracuse, New York, supplies crystal heads and controllers; SRS Stanford Research Systems, supplies low-cost QCM controllers; ColnaTec (www.colnatec.com), Gilbert, AZ, supplies sensor heads including a 24-crystal head; Tangidyne (tangidyne.us), formerly Virginia Sensors, supplies sensor heads, including a high-temperature (up to 900°C) sensor head that uses a gallium orthophosphate crystal; Intellimetrics (www.intellimetrics.com); Ulvac (www.ulvac-vacuum.com).
- Chopped ion gauge monitors: no longer available or supported, but many are in use using newer greases on the bearings.
- EIES: Inficon (www.inficon.com), East Syracuse, New York, supplies the EIES system Guardian, formerly sold as Sentinel.

Table 8.1 Comparison table of the deposition vapor sensors

Sensor	Measures	Characteristic	Sensitivity	Sampling rate
QCM	Flux	Material non-specific Intrusive	0.001 nm/s	1–10 Hz
Ion gauge	Vapor density	Material non-specific Intrusive	0.0001 nm/s	kHz
QMS	Vapor density	Material specific Intrusive	0.0001 nm/s	kHz
EIES	Vapor density	Material (atomic) specific Intrusive	0.0001–0.01 nm/s depending on the element	10 Hz
HCL-AA	Vapor density	Material (atomic) specific Non-intrusive	0.01–10 nm/s (depends on the atomic absorption line)	1–10 Hz
TDL-AAS	Vapor density and velocity (hence can obtain flux)	Material (atomic) specific Non-intrusive	0.0001 nm/s	10 Hz

- QMS: Hiden Analytical Ltd (www.hidenanalytical.com), UK, supplies QMS systems designed for vapor monitoring and source control, control of up to 16 signals.
- Atomic absorption: SVTA (www.svta.com), Eden Prairie, MN, supplies an HCL-AA system, AccuFlux Process Controller. A unique optical-electronic design is claimed to correct for drift. UV (~200 nm) optics is available as needed for atomic Se, Te, Sb, and As. The system originally developed by Lu and Guan, described in section 8.5.1 was sold commercially under the trade name 'Atomicas' for over a decade. However, the instrument is no longer manufactured or supported. TDL-AAS systems are not available commercially, but TDLs are available for a number of wavelength ranges from the following suppliers, among others: New Focus (www.newfocus.com), Santa Clara, CA; Toptica (www.toptica.com), Graefelfing (Munich), Germany.

8.7 Case studies

We discuss some specific thin film processes and their rate monitoring implementations.

8.7.1 MBE growth of oxide films with AA rate monitoring

MBE is a very precise evaporation technique usually practiced in ultra-high vacuum, although more recently oxides and nitrides are deposited by MBE at slightly elevated pressures of activated gas species. In order to achieve very accurate control of deposition rates one usually employs effusion cells that have very stable rates due to a stable temperature in the cell. Even so, deposition rates drop over time as the source material is depleted. To compensate for these slow changes in evaporation rates one can periodically adjust the source temperature (typically raise it) by doing a calibration. Conventionally such a calibration is commonly done with a simple ion gauge that is placed in the sample position. Typically an MBE system is equipped with shutters for each source, so each source can be calibrated separately. However, this requires moving the sample out while the calibration is done and the process is time consuming.

Several research groups have implemented their own versions of an HCL-AA system. As already discussed Klausmeier-Brown *et al.* and Chalmers and Killeen were the first to report such implementations for their MBE systems. More recently the Santa Barbara group (Jackson *et al.*, 1999) has reported improvements in reduction of long-term drift by additional chopping of the atomic beam at low frequencies. Note that the shuttered growth in MBE is ideal for AA as one can continuously correct for baseline drifts during periods when there is no flux. Bozovic and Matijasevic (2000) implemented a similar system in the Oxxel MBE that is now used at Brookhaven National Laboratory. Typical rates are very low in MBE deposition so one is not so concerned about a constant rate as for the integrated flux, or thickness of deposit. Thus the AA monitor signal can be integrated and the shutter can be controlled to attain a precise thickness of the deposit in each layer.

8.7.2 Co-deposition of Y, Ba and Cu during reactive co-evaporation of $YBa_2Cu_3O_y$

Much research has been devoted in the last two decades to making high-quality films of $YBa_2Cu_3O_y$ (YBCO) high-temperature superconductor; see for example the review by Matijasevic and Bozovic (1996). YBCO material is considered to be the most practical for many applications of high temperature superconductivity. Of the various approaches for making YBCO thin films, one particularly attractive and inexpensive method is co-evaporation from elemental sources. This method has three major challenges: uniform large-area heating, relatively high oxygen pressure during evaporation, and accurate atomic flux control. The large-area uniform heating and high oxygen pressure have been resolved rather elegantly with a method developed by Kinder and coworkers with a mechanical rotation of the substrate that allows for a black body heater with an oxygen pocket to be used (Berberich *et al.*, 1993). As for the elemental flux control several approaches have been employed.

The first and still most widely used method is to use QCMs with 'collimators', or tubes, in front of them (Kinder *et al.*, 1997). Collimators can select the source for the vapor to sample and block out the adjacent sources. This method works very well in high vacuum. However, at pressures above 10^{-5} torr there is a significant pressure effect. Figure 8.11 shows the pressure dependence of the QCM with a collimator. The comparison is for a QCM that is held at the sample position and has no collimator as pictured in Fig. 8.11, while only one source is used for evaporation. There is a fairly significant effect in flux for the pressure of interest in this process, about 5×10^{-5} torr. This effect is explained by scattering from the background gas. Some atoms are scattered away from the control QCM but cannot be scattered back due to the collimator, whereas atoms can scatter into the sample QCM. An additional problem is that each atomic species pressure effect is somewhat different and hence the atomic ratio will change as the pressure varies. The pressure effect can be mitigated if one can maintain the pressure in the chamber exactly during deposition and from run to run. However, maintaining a constant pressure is difficult to do in practice for this process since there are a number of different pressure regions in the system that all need to be kept constant.

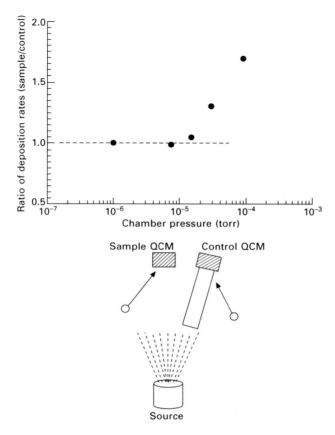

8.11 Pressure dependence of the control QCM for evaporation from a Cu source. The schematic shows the scattering effect presumed to cause the pressure dependence. Atoms can scatter into the sample QCM from higher angles, but they cannot scatter into the control QCM due to the collimator.

During the 1990s as atomic absorption sensors were being developed, both based on HCL and TDL, as described in Section 8.5, there was an industrial effort as well to implement these new sensors in manufacturing of YBCO wafers for microwave applications (Matijasevic and Slycke, 1998). This resulted in detailed comparisons of the various sensors that can be used in this process. Figure 8.12 shows a comparison of the pressure dependences for these sensors. QCMs were used with collimators. For the TDL-AAS sensor, a measurement of vapor density and velocity was used to yield a flux. Only the TDL-AAS flux sensor was deemed to be capable of attaining 1% composition control at pressures of 5×10^{-5} torr.

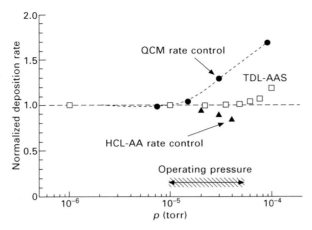

8.12 Chamber pressure dependence of the QCM, HCL-AA and TDL-AAS sensors.

8.7.3 Flux monitoring of atomic oxygen and nitrogen

Increasingly, the need for beams of atomic oxygen and nitrogen (AO/AN) to grow high-quality compounds is being recognized. Atomic absorption spectroscopy is a well-established method for the detection of atomic species in the upper atmosphere between 50 and 200 km (Iwagami et al., 2003). The advanced AO/AN sensor in development by Resonance Ltd departs from the usual fluorescence method to improve the measurement accuracy through a range of fluxes and vacuum conditions. The system optics operate entirely within the UHV chamber. VUV light from a modulated resonance lamp is filtered by a monochromator and detected by phase-locked solar-blind diodes. An open frame approach to the system makes it possible to measure atomic species within 2 cm of the substrate. A miniaturized VUV monochromator isolates the atomic resonance lines. The atomic beam actually passes through one arm of the open-frame monochromator. These features enhance accuracy by eliminating signals from other VUV emission sources, eliminating signals from unabsorbed lines, and by measuring the atomic beam *in situ* of the target-substrate.

A block diagram of the new instrument is pictured in Fig. 8.13. Stability of the nitrogen or oxygen resonance lamp at 120 or 130 nm is maintained by using thermal decomposition sources to maintain ~ 1 microbar O_2 or N_2 in the interchangeable lamp bulbs. A modulated 100 MHz RF drive dissociates the molecular species and excites the resonance lines. The bulb, grating, focusing mirror, and detector form a system that selects an atomic absorption feature and a second non-absorbed reference line. The oxygen reference line is an atomic line doublet that has a lower state at an excited level of O (135.7). This doublet is absorbed by O_2 but not by ground state O. Thus it can be

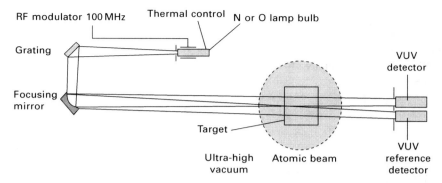

8.13 Block diagram of the resonance AO/AN-UHV atomic absorption system for *in situ* measurement of atomic species in MBE systems.

useful as an O_2 monitor and also a check on the short-term stability of the lamp. This line is about 5.6 nm from the 130.3 triplet used to monitor atomic oxygen. The nitrogen reference line is at 126.6 and like the O 135.7 line is not absorbed by the ground state of N. The detector signals are monitored by a data acquisition system for flexible display, analysis and storage of data.

8.8   Conclusions

This chapter has elaborated on various *in situ* deposition flux monitors and explored several specific examples of applications. However, the discussion is not exhaustive by any means and topics naturally came from the authors' greater areas of knowledge. We attempted to cover topics that have a sufficiently wide interest, but we also wanted to add certain material that complements already existing literature. It is our hope that this chapter gives the reader sufficient background to explore each topic further.

The current state of commercial vapor flux monitors provides instruments utilizing quartz crystal microbalances (QCM), quadrupole mass spectroscopy (QMS), electron-impact emission spectroscopy (EIES) and hollow-cathode lamp atomic absorption (HCL-AA). These instruments are available at reasonable cost and with different levels of sophistication for controlling deposition sources. Tunable-diode laser systems for AA vapor monitoring, inherently more powerful than HCL-AA, were fabricated in the past two decades on a custom basis and it is our hope that they will be available commercially in the near future. However, these systems are still very expensive and this has inhibited their broader development.

Speaking to the future of rate monitoring and control, the need for better rate monitoring ranges from its use in research and development systems and increasingly for large-scale high-rate and large-area processes that require better compositional control. The lack of adequate process rate monitors has impeded the advancement of such processes in research and in manufacturing. Very high rate processes often require electron beam co-evaporation from multiple sources that require special techniques to be successful in producing controlled composition over large areas. For example, the use of computer-controlled tomography combined with monitors that sample paths and/or points below a large-area substrate in numerous directions and/or points could control many sources to provide a uniform flux of a number of different elements over large deposition areas. We feel that these types of systems are technologically possible and could be available for application in the near future.

8.9 Acknowledgments

We would like to acknowledge Larry Lu of C. Lu Laboratory for extensive discussions on the topic of deposition rate monitors. Larry Lu has been an immense resource for us over the years and we thank him for that. We also thank Dr Hideki Yamamoto (NTT) for discussions of the EIES. We further thank Weizhi Wang of New Focus, and Hiden Ltd, SVTA, and Resonance Ltd for sharing information for this chapter. We are grateful to Larry Lu and Jeffrey Willis (LANL) for their proofreading of the manuscript. VM gratefully acknowledges the support from the Department of Energy Office of Electricity.

8.10 References

Appelboom, H., Matijasevic, V., Mathu, F., Rietveld, G., Anczykowski, B., Peterse, W., Tuinstra, F., Mooij, J., Sloof, W. & Rijken, H. 1993. Sm-Ba-Cu-O films grown at low temperature and pressure. *Physica C Superconductivity*, **214**, 323–334.

Benerofe, S., Ahn, C., Wang, M., Kihlstrom, K., Do, K., Arnason, S., Fejer, M., Geballe, T., Beasley, M. & Hammond, R. 1994. Dual beam atomic absorption spectroscopy for controlling thin film deposition rates. *Journal of Vacuum Science and Technology – Section B – Microelectronics Nanometer Structure*, **12**, 1217–1220.

Berberich, P., Assmann, W., Prusseit, W., Utz, B. & Kinder, H. 1993. Large area deposition of $YBa_2Cu_3O_7$ films by thermal co-evaporation. *Journal of Alloys and Compounds*, **195**, 271–274.

Bozovic, I. & Matijasevic, V. 2000. COMBE: a powerful new tool for materials science. *Materials Science Forum*, **352**, 1–8.

Buzea, C. & Robbie, K. 2005. State of the art in thin film thickness and deposition rate monitoring sensors. *Reports on Progress in Physics*, **68**, 385.

Campion, R., Foxon, C. & Bresnahan, R. 2010. Modulated beam mass spectrometer studies of a Mark V Veeco cracker. *Journal of Vacuum Science and Technology B*, **28**, C3F1.

Chalmers, S. & Killeen, K. 1993. Real-time control of molecular beam epitaxy by optical-based flux monitoring. *Applied Physics Letters*, **63**, 3131–3133.

Fazekas, E., Mezey, M. & Bunshah, R. 1972. Application possibilities of atomic resonance-absorption spectroscopy in vacuum metallurgy. *Journal of Vacuum Science and Technology*, **9**, 1373–1376.

Gogol, C. & Reagan, S. 1983. A performance comparison of vacuum deposition monitors employing atomic-absorption (AA) and electron-impact emission spectroscopy (EIES). *Journal of Vacuum Science and Technology A*, **1**, 252–256.

Hammond, R. 1975. Electron-beam evaporation synthesis of A15 superconducting compounds – Accomplishments and prospects. *IEEE Transactions on Magentics*, **MAG-11**, 201–207.

Hammond, R. 1978. Synthesis and physical-properties of superconducting compound films formed by electron-beam codeposition of elements. *Journal of Vacuum Science and Technology*, **15**, 382–385.

Iwagami, N., Shibaki, T., Suzuki, T., Sekiguchi, H., Takegawa, N. & Morrow, W. H. 2003. Rocket observations of atomic oxygen density and airglow emission rate in the WAVE2000 campaign. *Journal of Atmospheric and Solar-Terrestrial Physics*, **65**, 1346–1360.

Jaccard, Y., Cretton, A., Williams, E. & Locquet, J. 1994. Characterization of MBE-grown ultrathin films in the LaSrCuO system. *Proceedings of SPIE*, **2158**, 200–210.

Jackson, A., Pinsukanjana, P., Gossard, A. & Coldren, L. 1999. Noise, drift, and calibration in optical flux monitoring for MBE. *Journal of Crystal Growth*, **201**, 17–21.

Kieffer, L. & Dunn, G. 1966. Electron impact ionization cross-section data for atoms, atomic ions, and diatomic molecules: I. Experimental data. *Reviews of Modern Physics*, **38**, 1–35.

Kinder, H., Berberich, P., Prusseit, W., Rieder-Zecha, S., Semerad, R. & Utz, B. 1997. YBCO film deposition on very large areas up to $20 \times 20\,cm^2$. *Physica C: Superconductivity and its Applications*, **282**, 107–110.

Klausmeier-Brown, M., Eckstein, J., Bozovic, I. & Virshup, G. 1992. Accurate measurement of atomic beam flux by pseudo-double-beam atomic absorption spectroscopy for growth of thin-film oxide superconductors. *Applied Physics Letters*, **60**, 657–659.

Lu, C. & Guan, Y. 1995. Improved method of nonintrusive deposition rate monitoring by atomic absorption spectroscopy for physical vapor deposition processes. *Journal of Vacuum Science & Technology A: Vacuum, Surfaces, and Films*, **13**, 1797.

Lu, C. & Lewis, O. 1972. Investigation of film-thickness determination by oscillating quartz resonators with large mass load. *Journal of Applied Physics*, **43**, 4385–4390.

Lu, C., Lightner, M. & Gogol, C. 1977. Rate controlling and composition analysis of alloy deposition processes by electron-impact-emission-spectroscopy (EIES). *Journal of Vacuum Science and Technology*, **14**, 103–107.

Lu, C., Missert, N., Mooij, J., Rosenthal, P., Matijsevic, V., Beasley, M. R. & Hammond, R. H. 1989. Rate and composition control by atomic absorption spectroscopy for the coevaporation of high T superconducting films. *AIP Conference Proceedings*, **182**, 163–171.

Lu, C., Blissett, C. D. & Diehl, G. 2008. An electron impact emission spectroscopy flux sensor for monitoring deposition rate at high background gas pressure with improved accuracy. *Journal of Vacuum Science and Technology A*, **26**, 956–960.

Matijasevic, V. & Bozovic, I. 1996. Thin film processes for high-temperature superconductors. *Current Opinion in Solid State and Materials Science*, **1**, 47–53.

Matijasevic, V. & Slycke, P. 1998. Reactive evaporation technology for fabrication of YBCO wafers for microwave applications. *Proceedings of SPIE*, **3481**, 190.

Matijasevic, V., Garwin, E. & Hammond, R. 1990. Atomic oxygen detection by a silver-coated quartz deposition monitor. *Review of Scientific Instruments*, **61**, 1747–1749.
Missert, N., Hammond, R., Mooij, J., Matijasevic, V., Rosenthal, P., Geballe, T., Kapitulnik, A., Beasley, M., Laderman, S., Lu, C., Garwin, E. & Barton, R. 1989. *In situ* growth of superconducting YBaCuO using reactive electron-beam coevaporation. *IEEE Transactions on Magentics*, **25**, 2418.
Sauerbrey, G. 1959. Verwendung von Schwingquarzen zur Wägung dünner Schichten und zur Mikrowägung. *Zeitxhright für Physik*, **155**, 206–222.
Schellingerhout, A., Janocko, M., Klapwijk, T. & Mooij, J. 1989. Rate control for electron gun evaporation. *Review of Scientific Instruments*, **60**, 1177–1183.
Shockley, W., Curran, D. & Koneval, D. J. 1963. Energy trapping and related studies of multiple electrode filter crystals. *Proceedings of the 17th Annual Symposium on Frequency Control*, 88–126.
Smith, D. 1995. *Thin-film Deposition*, New York, McGraw-Hill.
Stirling, A. & Westwood, W. 1969. Investigation of the sputtering of aluminum using atomic-absorption spectroscopy. *Journal of Applied Physics*, **41**, 742–748.
Wajid, A. 1993. Long-life quartz crystals for dielectric coatings. *Surface and Coatings Technology*, **62**, 691–696.
Wajid, A. 1996. Method and a simple apparatus for rapid simultaneous measurement of resonance frequency and Q factor of a quartz crystal. *Review of Scientific Instruments*, **67**, 1961–1964.
Wajid, A. 1997. On the accuracy of the quartz-crystal microbalance (QCM) in thin-film depositions. *Sensors and Actuators A: Physical*, **63**, 41–46.
Wang, W., Hammond, R., Fejer, M., Ahn, C., Beasley, M., Levenson, M. & Bortz, M. 1995a. Diode-laser-based atomic-absorption monitor using frequency-modulation spectroscopy for physical vapor-deposition process-control. *Applied Physics Letters*, **67**, 1375–1377.
Wang, W., Hammond, R., Fejer, M., Ahn, C., Beasley, M., Levenson, M. & Bortz, M. 1995b. Diode-laser-based atomic absorption monitor using frequency-modulation spectroscopy for physical vapor deposition process control. *Applied Physics Letters*, **67**, 1375.
Wang, W., Fejer, M., Hammond, R., Beasley, M., Ahn, C., Bortz, M. & Day, T. 1996. Atomic absorption monitor for deposition process control of aluminum at 394 nm using frequency-doubled diode laser. *Applied Physics Letters*, **68**, 729–731.
Wang, W., Hammond, R., Fejer, M., Arnason, S., Beasley, M., Bortz, M. & Day, T. 1997. Direct atomic flux measurement of electron-beam evaporated yttrium with a diode-laser-based atomic absorption monitor at 668 nm. *Applied Physics Letters*, **71**, 31–33.
Wang, W., Hammond, R., Fejer, M. & Beasley, M. 1999. Atomic flux measurement by diode-laser-based atomic absorption spectroscopy. *Journal of Vacuum Science & Technology A: Vacuum, Surfaces, and Films*, **17**, 2676.
Wessel, G. 1949. Messung der Oszillatorenstärke der Bariumresonanzlinie. *Zeitschrift für Physik*, **126**, 440–449.

9
Real-time studies of epitaxial film growth using surface X-ray diffraction (SXRD)

G. ERES, J. Z. TISCHLER, C. M. ROULEAU, B. C. LARSON and H. M. CHRISTEN, Oak Ridge National Laboratory, USA and P. ZSCHACK, Argonne National Laboratory, USA

Abstract: We review recent results in the study of pulsed laser deposition growth kinetics using real-time surface X-ray diffraction. Interlayer transport as the primary driving force behind formation of atomically sharp layers is analyzed quantitatively from the measurements of time constants and shot-to-shot changes in single laser shot time dependent coverages in the growth of the model perovskite $SrTiO_3$. The results show that direct deposition into the open layers and very fast interlayer transport driven by energetic species during the arrival of the laser plume are the main components of layer growth per laser shot in both homo- and heteroepitaxy of complex oxides.

Key words: pulsed laser deposition, real-time surface X-ray diffraction, interlayer transport, single shot transient, perovskite $SrTiO_3$.

9.1 Introduction

The interface between two dissimilar materials often gives rise to exotic behavior and novel physical phenomena (Sutton and Balluffi, 1995). The family of complex oxides is a class of materials that has drawn great attention in the past decade because of the rich variety of physical phenomena that have been observed at interfaces between two oxides (Dagotto, 2007; Mannhart and Schlom, 2010; Zubko et al., 2011). The most widely studied example is the interface between $SrTiO_3$ (STO) and $LaAlO_3$ (LAO) (Ohtomo and Hwang, 2004). It is remarkable that phenomena such as two-dimensional conductivity, magnetism, and superconductivity can arise from an interface between two oxides that are insulating in bulk (Huijben et al., 2009; Reiner et al., 2009). The chemical abruptness and the crystalline perfection of the layers play a critical role in the manifestation of interfacial phenomena. The interface sharpness is critical for the manifestation of interface phenomena in full extent and it is important for both technological applications and for facilitating fundamental understanding of the phenomena in terms of simple structural models.

Epitaxial growth is a highly controllable method for systematically assembling dissimilar materials into artificial structures with atomic-scale precision (Schlom et al., 2008). The two most widely used techniques for epitaxial growth of complex oxide films are molecular beam epitaxy (MBE) (Schlom et al., 2001; Doolittle et al., 2005) and pulsed laser deposition (PLD) (Lowndes et al., 1996). Both methods are capable of controlling the layer thickness with submonolayer precision to produce interfaces that change from one oxide to the other over a single unit cell distance (Lee et al., 2005; Jang et al., 2011). For its ability to control the layer thickness and layer stacking in superlattice growth, MBE has been held as the standard for epitaxial growth precision (Farrow, 1995). Over the past few years the perceived reliability gap between MBE and PLD has been closed by advances in practical implementation of PLD and fundamental understanding of the mechanisms of PLD related to layer-by-layer (LBL) growth (Willmott and Huber, 2000; Willmott, 2004; Christen and Eres, 2008).

In this chapter we discuss the advances in fundamental understanding of the growth kinetics of PLD realized by real-time studies using surface X-ray diffraction (SXRD). We discuss the reasons why these advances have been slow to materialize in application of PLD, and conclude that the potential of PLD for controlling oxide film growth is still far from being fully utilized. The goal is to understand PLD growth kinetics in terms of atomic surface transport processes from time-resolved SXRD measurements, and is not intended to be a comprehensive review of the literature. The chapter is focused on the special case of homoepitaxy of STO as a model system to study the growth kinetics of the family of perovskites in its pure form unobscured by various thermodynamic factors such as surface energies, lattice strain, and thermal expansion mismatch (Gorbenko et al., 2002; Tromp and Hannon, 2002). The general principles derived from these kinetics studies are expected to be the foundation for understanding and controlling heteroepitaxial growth that is a critical factor in developing electronic devices based on the properties of complex oxides.

9.1.1 The general features of growth intensity oscillations

We start by describing the features of growth intensity oscillations that are in general common to all diffraction methods. The specific difference between particular techniques using X-rays, helium scattering, and electrons is mainly in the surface penetration depth that must be taken into account in quantitative data analysis (Zangwill, 1988; Woodruff and Delchar, 1994). Because we use SXRD for the measurements, the rest of this discussion refers to X-rays unless otherwise noted. For the purpose of analysis the measured diffracted intensity consists of three components:

$$I(q) = I_{\text{coh}}(q) + I_{\text{diff}}(q) + I_{\text{bkg}},\qquad [9.1]$$

where $I_{\text{coh}}(q)$ refers to the sharp specular peak that describes long-range order, $I_{\text{diff}}(q)$ corresponds to the broad diffuse background surrounding the central peak that describes the short-range correlations, I_{bkg} is a weak slowly varying background component that is in most cases negligible. The vast majority of kinetic studies have been performed by measuring the intensity of the sharp central peak (Vlieg et al., 1988, 1995; Fuoss et al., 1992; van der Vegt et al., 1992). It is important to note that the separation of $I_{\text{coh}}(q)$ + $I_{\text{diff}}(q)$ depends delicately on the growth mode, and more significantly on the quality of film growth. Therefore, the analysis of the film growth kinetics in terms of layer coverages is 'cleaner' the closer the system is to LBL growth mode. High-quality growth is particularly important in PLD because it allows measurement of time-dependent coverages at the single laser shot level (Eres et al., 2002; Tischler et al., 2006; Christen and Eres, 2008). Analysis of the time-dependent coverages as a function of growth conditions enables extracting the surface transport timescales and narrowing down the possible mechanism of PLD growth.

The relationship between the intensity of the central peak and the coverage is according to the kinematic approximation described by the following equation:

$$I(q) = |F_c(\text{HKL})|^2 |(\theta_0 - \theta_1) - (\theta_1 - \theta_2) + (\theta_2 - \theta_3) - \ldots|^2 \qquad [9.2]$$

where F_c is the structure factor, $(\theta_n - \theta_{n+1})$ for $n = 0, 1, 2 \ldots$, is the exposed coverage, and $\theta_0 = 1$ for the substrate. For the special case of only two levels ($n = 0, 1$) islands on top of the substrate Eq [9.2] describes LBL growth. The intensity of the central peak of a particular reflection is measured using a point detector. For cubic lattices the measurement point is typically the (0 0 ½) anti-Brag position where the surface sensitivity is the highest. When the X-ray beam is aligned at the destructive interference position the intensity is fully attenuated at $\theta = 0.5$ and it recovers to its initial value again at a full layer. In this configuration the intensity of the specular rod describes the distribution of the growth species in layers parallel to the growing surface. The specular intensity contains no information concerning the size, shape, and the spatial distribution of islands. As the coverage increases from 0 to 1 in filling up of successive layers, the intensity undergoes periodic oscillations described by:

$$I(\theta) = (1 - 2\theta)^2, \qquad [9.3]$$

with the period of one oscillation corresponding to the time that is needed for growing one full layer. The sharp parabolic cusps in these oscillations shown in Fig. 9.1(a) are the signatures of LBL growth.

Real growth systems are characterized by some degree of damping of the growth oscillation intensity envelope. A great deal of research has gone into

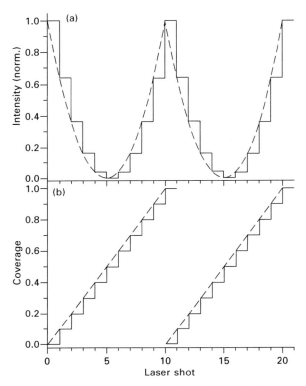

9.1 Comparison of the intensities in (a) and the coverages in (b) for LBL growth in continuous mode given by the dashed line, with pulsed LBL growth given by the solid lines.

generalizing these departures from LBL growth and relating them to actual growth parameters of the system. The most often observed departure is a decay of the intensity envelope that occurs as a consequence of the spreading of the growth front over multiple layers. At the simplest level of growth kinetics studies the growth intensity oscillations can be used to extract the layer coverages that provide a picture of how the interface formed and other parameters such as the interface width and surface roughness.

9.1.2 Simple models of interlayer transport

Interlayer transport is the migration of adatoms across step edges into a lower layer (Rosenfeld *et al.*, 1997; Michely and Krug, 2004). The probability of this transport is affected by a step edge barrier, the so-called Ehrlich–Schwoebel barrier that arises from incomplete coordination of atoms at the terrace edges (Ehrlich and Hudda, 1966; Schwoebel and Shipsey, 1966). In contrast to intralayer transport or surface diffusion that is migration on a flat terrace and corresponds to lateral transport, interlayer transport refers to the effectiveness of downward vertical transport on the growing surface (Zhang and Lagally, 1997). The connection between the diffracted intensity, coverage, and interlayer transport was first made by a formalism discussed by Cohen et al. (1989). In the simplest form of this picture for a given adatom flux F, the adatoms migrate from the top of the growing islands into lower layers with an interlayer transport rate constant of k. The pulsed nature of PLD enables time-dependent studies of the growth process and the separation and identification of the various surface transport processes and the determination of their time constants (Karl and Stritzker, 1992; Chern et al., 1993; Blank et al., 1999; Lippmaa et al., 2000; Eres et al., 2002; Fleet et al., 2005). Interlayer transport plays a central role in PLD and to recognize its time-dependent features in actual growth we first discuss a couple of simple model systems.

The simplest possible model system is pulsed LBL growth, illustrated in Fig. 9.1(a), that has interlayer transport as the only transport process. The first notable feature of pulsed growth is that it breaks up the continuous parabolic growth intensity oscillation into discrete steps. The number of steps is $1/p$, where p = the number of pulses that are needed for growing one full layer. The key assumption in pulsed LBL growth is that interlayer transport is infinitely fast. This assumption leads to the discrete drops and jumps in the intensity that correspond directly to the coverage changes. The perfectly flat segment separating two successive pulses is indicative of the absence of time-dependent transport processes in growth. Infinitely fast interlayer transport leads to the extreme situation that all adatoms go into the topmost unfilled layer until it is complete. This growth mode in which the growing layer is complete before a new layer starts growing is referred to as LBL growth. Such a system cannot exist on a vicinal surface because step flow would supersede LBL growth when the characteristic length scale of a terrace size is reached (Rosenfeld et al., 1997; Michely and Krug, 2004).

The next model system closer to real growth is one that has a finite interlayer transfer rate. For a system of islands on a flat surface (a two-level system) the intensity dependence on the coverage can be calculated according to the transport model by Cohen et al. (1989). In this system the intensity is allowed to recover fully to its maximum possible value for a particular step (see Fig. 9.2b) before the next pulse is applied. The time-dependent coverage in the recovery is obtained by solving the simple differential equation:

$$d\theta_2/dt = -k\theta_2(1 - \theta_1) \qquad [9.4]$$

to get the intensity as:

$$I(t) = [2b - 1 + 4\theta_2(t)]^2 \qquad [9.5]$$

where k is the interlayer transport rate constant and b is an integration

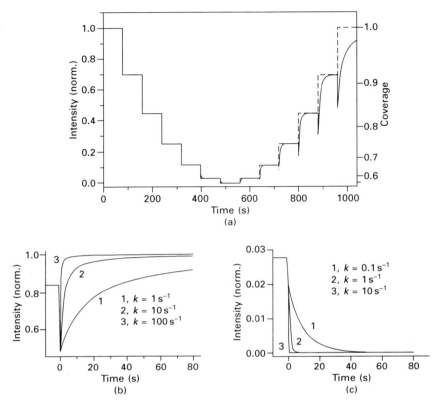

9.2 A finite interlayer transport rate shown by the solid line in (a) is manifested as recovery toward the infinitely fast interlayer transport steps given by the dashed line. The plots in (b) show that the change of the interlayer transport rate causes large changes in the shape of the recovery curves near full coverage, and barely observable rounding off in (c) near and below half coverage.

constant given by the initial coverage θ_1. In Fig. 9.2(a) the intensity recovery at a given initial coverage (θ_1) for a fixed coverage increase $\Delta\theta = 1/p$ is compared with pulsed LBL growth. The first important feature of note is that for a finite value of interlayer transport rate the intensity curve is no longer symmetric around the minimum intensity point ($\theta = 0.5$). This asymmetry is the manifestation of the fact that at low coverages interlayer transport is more effective than at high coverages, the island sizes are small and the adatoms can transfer off the top of the islands very quickly. With increasing coverage there are fewer holes available and the adatoms must travel longer to find them. Figure 9.2(b) shows that increasing the interlayer transport rate – this is equivalent to increasing the substrate temperature in growth – speeds up the recovery, and for very fast interlayer transport the intensity curves approximate the discrete steps in pulsed LBL growth. It is also important to note that for sluggish and incomplete interlayer transport the segment separating two successive pulses is no longer flat. Because of the quadratic dependence of the intensity on the coverage, the diffraction signal is most sensitive to interlayer transport near full coverage (Ichimiya and Cohen, 2004). The important conclusion from analyzing this simple model with finite interlayer transport is that the appearance of the recovery signal is affected not just by the time constant $k = 1/\tau$, but also by the initial coverage (see Figs 9.2b and 9.2c) where the coverage change occurs.

9.2 Growth kinetics studies of pulsed laser deposition (PLD) using surface X-ray diffraction (SXRD)

9.2.1 Reasons for using X-rays

The purpose of using diffraction techniques in film growth studies is to obtain surface-specific real-time information on the formation and evolution of the crystalline structure on the growing surface. Over the past two decades *in situ* monitoring in epitaxial film growth has become synonymous with reflection high-energy electron diffraction (RHEED) (for details see Chapter 1 by Koster) (Braun, 1999; Ichimiya and Cohen, 2004; Rijnders and Blank, 2005). The use of RHEED as a growth monitoring tool is widespread primarily because the equipment is affordable, relatively easy to implement, and valuable information on the surface structure and surface ordering can be obtained instantaneously by qualitative examination of the diffraction patterns (Ingle *et al.*, 2010). The grazing incidence geometry of RHEED provides a high degree of surface sensitivity and the configuration of the source and detection scheme does not interfere with the growth environment. However, the strong interaction of electrons with the surface causes dynamic scattering effects that require a difficult and complex theoretical treatment to explain the results. The most important feature of RHEED for growth kinetic studies is the periodic oscillation of the intensity of the diffraction spots with increasing layer thickness. Despite widespread use, the origin of the RHEED intensity oscillations remains controversial and the contributions to the scattering intensity by the coverage and the step edge density are still unresolved (Clarke and Vvedensky, 1987; Braun *et al.*, 1998; Korte and Maksym, 1997; Joyce and Joyce, 2004). The complicated data analysis that is required to treat the RHEED scattering intensities limits RHEED to a qualitative role in growth kinetics studies.

X-ray diffraction has been the dominant tool for bulk crystallography since the discovery of X-rays (Giacovazzo *et al.*, 2011). The superior ability of X-rays to measure length scales precisely is of great significance for the study of perovskite interfaces in film growth because their properties can be dramatically affected by minute changes in lattice parameters, atomic coordination, and symmetry. However, the large penetration depth and the low signal intensity make X-ray diffraction unfavorable for surface studies. The 10^{12} photons/s intensity available at third generation synchrotrons helps to overcome the main obstacle for application of X-rays in surface structure studies and opens up the possibility of using X-ray diffraction as a routine surface science tool. These high intensities enable the use of special configurations and techniques that maximize the surface sensitivity of X-rays leading to rapid growth of SXRD applications. The increasing availability of beam time at third generation synchrotrons over the past decade has resulted in the rapid growth of techniques using X-ray diffraction for surface studies. Further details of performing SXRD with synchrotron X-rays will be given in the following section.

The main justification for using synchrotron X-ray diffraction for growth kinetics studies is the straightforward data analysis (Fuoss and Brennan, 1990; Braun and Ploog, 2006). The weak interaction of X-rays with the surface makes the kinematic treatment of the scattering intensity fully adequate for analyzing the data (Kaganer, 2007). The key advantage of SXRD is that, unlike RHEED, it enables quantitative determination of surface coverages directly from the measured intensities (Rijnders and Blank, 2005). The step edge density is irrelevant for SXRD. The analysis of the scattering intensity in terms of the coverages of incomplete layers is the starting point for testing and validating simple growth models. Another advantage related to the high brilliance of synchrotron X-rays is that time-resolved measurements of structural changes can be performed. The faster the timescales are, the higher is the X-ray intensity needed to perform the measurements with adequate signal to noise ratio. The processes that are most interesting for understanding the mechanism of PLD growth occur on a microsecond or faster timescale.

9.2.2 The use of SXRD for growth kinetic studies

The surface sensitivity of X-ray diffraction originates from the truncation of the three-dimensional periodicity of the bulk lattice by the surface. The X-ray scattering theory of the surface sensitivity has been treated extensively and for details the interested reader is referred to the books and a number of excellent review articles cited in the references (Feidenhans'l, 1989; Robinson and Tweet, 1992; Als-Nielsen and McMorrow, 2001). As a consequence of the abrupt termination of the crystal lattice, the Bragg points of the bulk lattice are connected by streaks of intensity with the minimum value at the mid-point (anti-Bragg) between two neighboring Bragg peaks, corresponding to scattering from a single lattice plane. The mathematical form of the scattering intensity is given by:

$$I \propto |F_c(HKL)|^2 |F_{ctr}(L)|^2 \text{ with } |F_{ctr}(L)|^2 = 1/[4\sin^2(\pi L)] \qquad [9.6]$$

where F_c the structure factor contains all the atomic coordinates. The rods of scattering intensity described by this equation are referred to as crystal truncation rods (CTRs). The CTRs provide information about the vertical structure of the surface and are extremely sensitive to changes of the lateral periodicity with respect to the bulk. Several orders of magnitude variation of the scattered intensity can be observed with changes in atomic site occupancy, surface relaxation, presence of adatoms, and surface roughness. Feidenhans'l, (1989) and Saldin and Shneerson (2008) are excellent review articles providing details concerning the use of CTRs for surface structure determination.

The most significant feature of the CTRs for growth kinetics studies is that they are sensitive to the surface coverage. This dependence of the scattering intensity on the coverage illustrated in Fig. 9.3 is used for performing growth kinetics studies by measuring the intensity of the CTRs at some point along the CTRs. The surface specificity is the highest at the anti-Bragg point. The perfection of the initial surface is a critical requirement for using the CTR intensity measurements for growth kinetic studies because surface roughness weakens the CTRs, resulting in significantly reduced scattering (Robinson, 1986).

9.2.3 Experimental set-up

The typical components of the experimental set-up for SXRD studies of film growth include a PLD chamber that is coupled or integrated with a

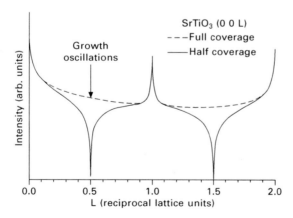

9.3 Illustration of the intensity of the (0 0 L) specular crystal truncation rod at half and full surface coverage. The anti-Bragg position marked with the arrow is where the largest intensity changes occur with coverage and is the most favorable point for measuring the growth intensity oscillations.

diffractometer used for positioning the sample and the detector with respect to a fixed incident X-ray beam. The chamber is equipped with two Be windows to permit the entrance and the exit of X-rays with minimal attenuation and scattering. Beyond the basic components, the actual implementation of SXRD measurements can vary greatly in terms of the chamber configuration, the diffractometer geometry, and the detection method and scheme chosen.

Our PLD chamber is mounted on a diffractometer that uses the (2+2)-circle diffraction geometry that is optimized for surface X-ray diffraction measurements (Evans-Lutterodt and Tang, 1995). The schematic illustration of the main components of this system is shown on Fig. 9.4 and a picture of the portable growth system in Fig. 9.5. The diffractometer has four degrees of freedom consisting of two angles γ and δ for setting the position of the detector, and χ and ϕ for setting the angle of incidence and the azimuth of the sample, respectively. The sample is mounted with its surface in the vertical plane and a set of three motorized micrometer pushrods attached to the back of the sample holder are used for changing the two tilt angles χ_1 and χ_2 that are used for aligning the surface normal perpendicular to the incident X-ray beam. The X-ray beam enters through a 2.5 cm (one inch) wide rectangular Be window and the diffracted X-rays exit through a recessed cylindrical segment Be window that covers an angle of 120° at 10 cm (4 inch) radius

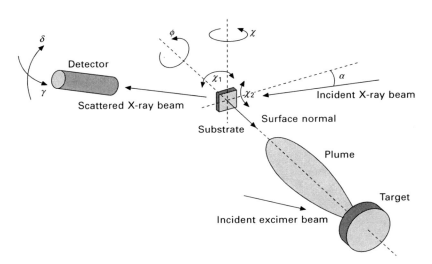

9.4 Schematic illustration of the (2+2)-circle diffractometer angles for real-time measurements of PLD growth by SXRD. The angles γ and δ are used to position the detector. The angle of incidence of the X-ray beam on the sample is set using χ, and ϕ changes the azimuth of the sample. The sample normal is aligned perpendicular to the incident X-ray beam using a set of three motorized micrometer pushrods that adjust the angles χ_1 and χ_2.

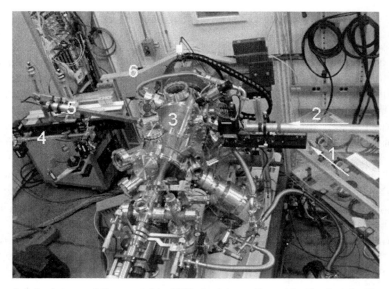

9.5 A picture of the portable PLD chamber aligned in the hutch. The excimer laser beam path is designated by 1, the X-ray beam enters the chamber along 2, and the scattered X-rays exit through a large Be window 3. The detector arm shows the bottom detector 4 aligned on the (0 0 ½) and the top detector 5 aligned on the (0 1 ½) position. The δ motion of the detector is shown by 6.

and is 12.5 cm (5 inches) wide. The scattered X-rays are detected by either point detectors of the thallium-doped NaI scintillator type or by fast avalanche photodiode detectors mounted on the detector arm. Using area detectors such as a CCD camera no scanning is required and the entire diffraction profile can be captured faster to enable time-resolved studies of the broad diffuse scattering around the CTR.

The excimer laser enters the chamber through a quartz window and a projection optical system is used for obtaining a laser spot with an exactly defined shape and size with a uniform energy density on the target. Laser ablation is performed in the typical fluence range of $1–3 \, J/cm^2$. The targets are loaded in a computer-controlled multi-target carousel that accommodates three targets, and is programmable for superlattice growth. The sample heating is performed by a pyrolytic boron nitride encapsulated graphite filament heater. The sample temperature is determined by measuring the lattice constant expansion at the (0 0 2) Bragg angle using a Ge analyzer and comparing it with the thermal expansion data for STO. The chamber is evacuated by a turbo molecular pump to a background pressure in the low 10^{-7} torr range. PLD of STO was performed in an oxygen pressure range from vacuum to 10 mtorr.

9.3 Real-time SXRD in SrTiO₃ PLD: an experimental case study

The preparation of the STO substrate surface is a critical step in the SXRD growth experiments. It is known that high-quality growth occurs only on TiO_2-terminated STO surfaces (Kawasaki *et al.*, 1994; Koster *et al.*, 1998; Ohnishi *et al.*, 2004). There is a large amount of literature regarding the exact details of how such surfaces are prepared by different groups. The previous studies of static STO surface structure by CTR measurements provide background information about the state of the starting STO surface (Charlton *et al.*, 2000; Herger *et al.*, 2007). The most important factor for SXRD is the initial scattering intensity by the substrate measured at the growth temperature. The initial intensity for good substrates at 650 °C is higher than 5×10^5 cps, and for excellent substrates exceed 10^6 cps and is roughly equal for the specular (0 0 ½) and the off-specular (0 1 ½) rod. The higher these initial intensity values are, the better the time-resolution that can be achieved and the faster the kinetics that can be studied.

The scattering conditions in the film growth experiments are set to monitor the formation of STO unit cells (u.c.). The scattered intensity is measured simultaneously at the (0 0 ½) specular and the (0 1 ½) off-specular rod before, during, and after each laser shot. The significance of measuring simultaneously the intensity of a specular and an off-specular rod is that the timescale of surface ordering can be confirmed from the time delay between the transients corresponding to the two rods. The specular rod has momentum transfer along the surface normal and provides information only about the arrival and height distribution of material. The in-plane registry with the lattice on the growing surface corresponding to crystal growth is determined by measuring the intensity of an off-specular rod (H,K) ≠ (0,0) which has an in-plane momentum transfer component.

Persistent growth intensity oscillations illustrated in Fig. 9.6(a) are observed simultaneously at both specular and off-specular rods during STO PLD from 310 to 780 °C (Eres *et al.*, 2002). Figures 9.6(b) and 9.6(c) show that the PLD growth oscillations made up of single laser shot signal retain the overall parabolic shape. In the rest of this chapter the steps corresponding to single laser shots are referred to as SXRD transients or just transients. Each SXRD transient is a unique description of the conditions on the growing surface for the time duration between two laser shots referred to as the dwell time. The transients have a characteristic shape that depends on the surface coverage at the time of the laser shot. All transients that occur below half-coverage have a different shape shown in Fig. 9.6(c) from those above half-coverage. The transients clearly consist of two stages as Fig. 9.6(d) shows that occur on vastly different timescale.

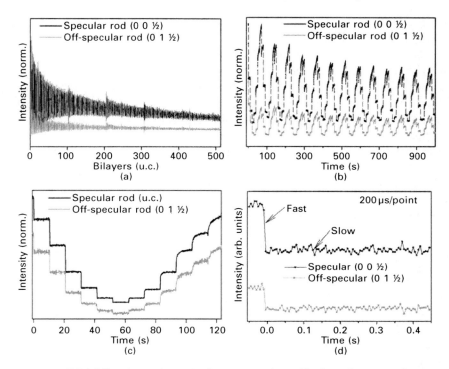

9.6 (a) Persistent layer-by-layer growth oscillations for more than 500 bilayers of STO simultaneously measured for both specular (black) and off-specular rods (gray). (b) Parabolic growth intensity oscillations for both (0 0 ½) and (0 1 ½) rods showing the single laser shot transients. (c) One full period showing time-resolved individual laser shots for both specular and off-specular rods. (d) Single shot transients measured with 200 μs/point time resolution clearly show the fast and the slow steps.

9.3.1 Surface transport timescales

In a continuous growth system such as MBE the coverage evolution is governed by a steady state among the various surface transport processes. Chopping a continuous beam was explored as a way of uncoupling the various surface transport processes and determining their role in the surface structure formation. The few experimental studies performed by chopping continuous molecular beams have produced inconclusive results on the benefits of incident flux modulation (Larsson *et al.*, 1995). The weak effect from chopped beams suggests that pulsing of thermal beams alone is not sufficient to produce smoother surfaces. Subsequent work directly comparing MBE and PLD growth concluded that the energetic effects of arriving growth species play an important role in the ability of PLD to produce a smooth surface (Taylor and Atwater, 1998; Hinnemann et al., 2001; Shin and Aziz, 2007; Warrender and Aziz, 2007, Aziz, 2008; Schmid et al., 2009).

The inherently pulsed nature and the enormous dynamic range of PLD with ratios of instantaneous to time-averaged growth rates that by some accounts exceed six orders of magnitude make PLD an ideal method for real-time studies of growth kinetics. These studies are greatly needed because the mechanism of PLD remains unresolved and the topic is not immune to occasional controversy (Tischler et al., 2006; Willmott et al., 2006; Vasco and Sacedón, 2007; Ferguson et al., 2009). In the prevailing picture of PLD the deposition and the growth processes are assumed to be two separate stages. The deposition stage consists of the landing of the laser plume on the growing surface where the growth species begin their search for the proper crystallographic sites. The growth stage is best described as the actual time spent on the growing surface before incorporation at the proper crystallographic site occurs. This is a very important timescale for understanding the PLD mechanisms because this is the stage during which the extra kinetic and internal energy that the growth species arrive with must be transformed or dissipated.

Some insight into the rate at which this extra kinetic energy is dissipated can be gained from theoretical calculations and Monte Carlo simulations of energetic ions impinging on the surface (Jacobsen et al., 1998; Adamovic et al., 2007). These calculations show that the energy of ions in a 20–50 eV range is dissipated in a few collisions that occur on a picosecond timescale. Obviously this timescale is inaccessible by any current diffraction technique. These simulations provide direct evidence that the transition from three-dimensional (3D) growth to two-dimensional growth (2D) is induced by atomistic processes that occur within the initial 10 ps following the collision with the growing surface (Adamovic et al., 2007).

9.3.2 A brief stage of perfect layer growth

A closer look at the features of the SXRD transients reveals that the growth of the first layer in PLD growth of STO is uniquely different from the rest of the growth. The kinetic signature of the SXRD transients in this regime shown in Fig. 9.6(c) is that of the sharp drops and flat steps that characteristically occur only for the model system of pulsed LBL growth in Fig 9.1(a). This is a special situation that occurs only on extremely high-quality substrates that have very few or no residual holes left by the substrate preparation step on the starting STO surface. On these perfectly terminated initial surfaces STO growth starts by nucleation of small islands of a few nm size. The average island spacing shown in Fig. 9.7(a) corresponds to a parameter referred to as the nucleation distance (ℓ) (Pimpinelli and Villan, 1998). At this stage the number and density of islands increase rapidly, but cannot increase indefinitely

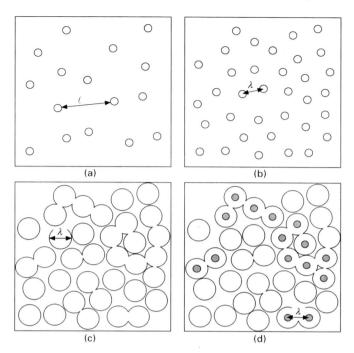

9.7 (a) The nucleation distance ℓ at low coverage. (b) The island separation λ near saturation island density. (c) The onset of coalescence occurs when λ matches the island diameter $2r$ resulting in formation of networks of interconnected islands. (d) Nucleation of a new top layer (black islands) on top of the base layer by the first shot after coalescence.

because as the inter-island spacing falls, the growth species are more likely to collide with the edges of existing islands, than with each other. This island density is known as the saturation island density (N_s) and is characterized by the smallest inter-island separation (λ) shown in Fig. 9.7(b) that can occur on a particular growing surface. PLD is unlike other film growth methods in that N_s, which typically occurs at a few percent of surface coverage, is almost always reached during the very first laser shot (Lam *et al.*, 2002; Schmid *et al.*, 2009). The growth on this surface is LBL growth because on smooth starting surface there can be only one layer open (incomplete) and all incoming material must go into this single open layer. As already mentioned, LBL growth is a hypothetical situation that is unsustainable in real systems (Rosenfeld *et al.*, 1997; Michely and Krug, 2004).

This brief stage of LBL-like growth ends when the island size (r) reaches λ as shown in Fig. 9.7(c). The mechanism by which r increases to reach λ is coalescence, which occurs when the islands grow so large that their edges run into each other (Zinke-Allmang, 1999; Evans *et al.*, 2006). In a post-coalescence regime a percolation network forms that corresponds to an instantaneous jump in the effective island size (or connectivity length) because the islands are partially or fully interconnected into a larger continuous network. The significance of r reaching or exceeding λ is that now the island nucleation requirement is satisfied on top of the growing islands and new islands must nucleate on the growing layer. The growing layer from here on is referred to as the 'base' layer to differentiate it from the substrate as a layer on which a second layer can nucleate, which is referred to as the 'top' layer. It is well established by mathematical models of various levels of sophistication and independently by experimental studies of numerous growth systems that the onset of percolation on surfaces always occurs before full coverage (Evans et al., 2006). The percolation threshold is 0.593 for a random adlayer on a square lattice and the onset can be delayed to about 0.8 for special island configurations. The experimental values estimated for the percolation threshold in STO PLD fall in the range from >0.5 to <0.8. The onset of coalescence in STO completes the brief stage of LBL-like growth and opens up a new growth stage that is the highest perfection growth mode physically possible by any growth method, and it is this growth mode that is truly representative of PLD film growth.

9.3.3 Simultaneous two-layer growth

Island coalescence transforms the growing surface from island growth on a flat surface to island growth on a base layer with holes (island growth on top of islands that are on a flat surface). This is a permanent transformation and a single layer growth surface (truly atomically flat surface) never appears again because the conditions for nucleation of the next layer are always satisfied before the previous (base) layer is completed. It is important to note that a step edge (Ehrlich–Schwoebel) barrier is not necessary for the nucleation of the second layer (Rosenfeld et al., 1997; Michely and Krug, 2004). However, the presence of a step edge barrier would accelerate the growth front broadening and force an earlier transition into three-dimensional growth (Tersoff et al., 1994). Coalescence is a convenient reference point for distinguishing the growth behavior that occurs simultaneously on the two respective layers. The island nucleation and island growth process on top of the base layer (top layer) is referred to as the pre-coalescence stage and the hole filling in the base layer as the post-coalescence stage. In this growth mode that we refer to as simultaneous two-layer (S2L) growth there is always one layer in the pre-coalescence and one layer in the post-coalescence stage of growth.

An important consequence of the opening of an extra layer is that the intensity maxima in the growth oscillations never again reach their initial value. The intensity maxima (marked by dashed lines in Fig. 9.8) now occur at coverage where more adatoms join islands on top of the base layer than are filling holes in the base layer (i.e. equal hole and island areas). The magnitude of these maxima is always less than the initial intensity because even when the base layer is complete there are already islands that according to Eq [9.2] interfere destructively with the base layer and reduce the intensity.

The simple model by Cohen discussed in Section 9.1.2 vividly illustrates the important role that interlayer transport plays in the formation of sharp interfaces. There is a striking similarity between the features of the PLD SXRD transients in Fig. 9.8 with the simple model systems in Section 9.1.2, suggesting that PLD is remarkably close to LBL growth. In practice, the Cohen-type transport models never work satisfactorily because a single interlayer transport is unable to capture the complexities of surface transport during film growth in real systems. The usual approach is to introduce fitting parameters to extract the surface coverage by fitting the growth intensity oscillations (van der Vegt *et al.*, 1992; Alvarez *et al.*, 1998). In this approach the true growth kinetic information is not available because the physical meaning of the fitting parameters is not clear. Nevertheless, the important contribution of the early SXRD studies of continuous growth systems (MBE) is that they show that a simple mathematical model can be used to unambiguously connect the diffracted intensity with coverage evolution that is governed by some type of surface transport.

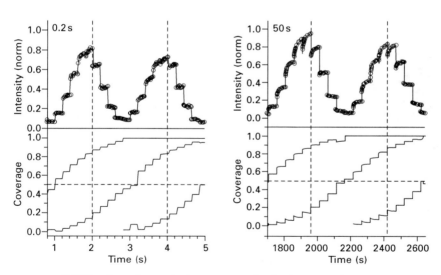

9.8 The time-dependent coverages in the bottom plots were extracted by calculating the intensity of the growth transients given by the solid black lines for each measured data point given by the circles for 0.2 s and 50 s dwell time. The vertical dashed lines mark one oscillation period as the time between two successive maxima, and the horizontal dashed lines mark the half coverage point.

256 *In situ* characterization of thin film growth

Instead of fitting the data to transport models we developed an approach that allows direct determination of the time-dependent surface coverages from the diffracted intensity (Tischler *et al.*, 2006) under two-layer growth conditions. This analysis is performed without any further assumption about the physics of the underlying growth process, such as the shape of the islands. The analysis is based on the fact that the kinematic approximation allows a straightforward calculation of the intensity subject only to the coverage of the incomplete layers. In contrast the RHEED intensities must be interpreted in terms of the step edge density.

The validity of the two-layer assumption for calculating the time-resolved coverage from the SXRD intensities was confirmed by analysis of the surface roughness of the starting growth surface and the films grown on the same substrates. AFM imaging of a great number of films illustrated in Fig. 9.9 with different thickness grown under different conditions shows that even for films more than 100 u.c. thick the surface roughness for high-quality S2L PLD growth never exceeds two layers. The best samples that exhibit persistent growth intensity oscillations show minimal growth front broadening compared to the substrate as shown by the histograms in Fig. 9.9(b). Because the overall intensity is the result of cancellation of the scattering contribution from all even layers by all odd layers (see Eq. 9.2), the interface width cannot be determined uniquely from the value of the intensity alone. The AFM images provide independent data that exclude the possibility of interface broadening beyond two layers in high quality S2L growth and confirm that

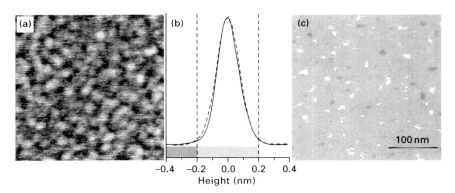

9.9 Atomic force microscopy (AFM) image of the STO film surface after growth of more than 20 u.c. given in standard image presentation in (a). The same image was plotted in (c) by adjusting the grayscale shades defined at the bottom of the histogram in (b) to change at one u.c. height to convey what the X-rays are 'seeing' and to aid visualization of the height distributions. The histogram in (b) compares the surface width after growth (dashed line) with the surface width of the starting surface of STO (solid line).

the data are consistent with a two-layer (three-level) surface illustrated in Fig. 9.10.

Accordingly, the starting point for the quantitative analysis of the SXRD transient intensities is given by the following two-layer form of Eq. [9.2]:

$$I(t) = I_0|(1 - 2\theta_1(t) + 2\theta_2(t)|^2 \qquad [9.7]$$

where I_0 represents the initial intensity and $\theta_1(t)$ and $\theta_2(t)$ are the fractional coverages of the base layer $\theta_1(t)$ and the top layer $\theta_2(t)$, respectively. With the constraint that the deposition per shot $1/p = \theta_1(t) + \theta_2(t)$ is fixed, the time-dependent coverages $\theta_1(t)$ and $\theta_2(t)$ are the solution of the above equation for $I(t)$ at each experimentally measured data point. It is important to understand that this approach does not force a fit to the starting structure of islands on an incomplete base layer. This method distinguishes directly between LBL growth and S2L growth through the initiation of coverage of a top layer before completion of filling of the base layer in order to account for the observed intensities as a function of deposition.

The single laser shot time-dependent (intra dwell-time) coverages of the individual layers extracted from the SXRD intensities by this two-layer approach are the most important data available from the STO PLD growth

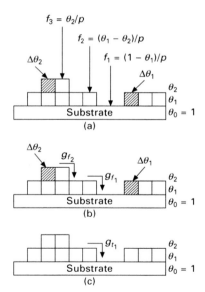

9.10 (a) The expected coverages for the three levels given by fractions f_n (n = 1–3) of the incoming plume distributed on the growing surface according to the exposed surface area. (b) The components of fast interlayer transport g_{fn} and the growth of the layer $\Delta\theta_n$. (c) Thermal interlayer transport is separated out to emphasize that STO growth is over and that only already crystallized STO transfers in this step.

experiments. The time-dependent coverages allow determination of the time constants and identification of the particular surface transport processes based on the magnitude and the dependence of the time constants on the growth parameters. The time-dependent coverages shown in Fig. 9.8 resemble a rising staircase. Comparison with the pulsed LBL growth model in Fig. 9.1(a) reveals that each step is unique and slightly different from the simple model. Similar to the simple models, each coverage step magnified in Fig. 9.11 has two components that are clearly separable; an instantaneous jump that corresponds to a very fast interlayer transport process, and a slow part that occurs during the dwell time between successive laser shots.

The subtle differences in the shape of the slow component are indicative of net interlayer transport for a particular shot. These steps can be categorized

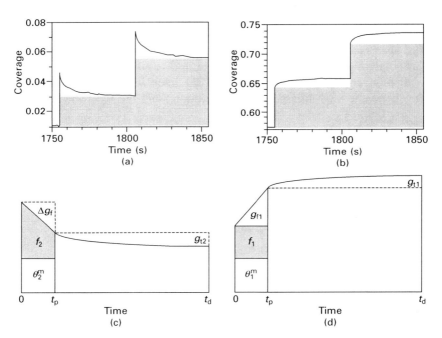

9.11 Pre-coalescence and post-coalesce time-dependent coverages (a) and (b), respectively replotted from Fig. 9.8 for 50 s dwell time show the departure from the shaded steps for pulsed LBL growth. (c) Schematic illustration of pre-coalescence components of interlayer transport: $t = 0$ expected coverage, actual coverage after fast interlayer transport Δg_f out of the layer is completed at $t = t_p$ (the time of plume arrival), thermal interlayer transport g_{t2} out of the layer is measured at $t = t_d$ (the dwell time). (d) Schematic illustration of post-coalescence components of interlayer transport: $t = 0$ expected coverage, actual coverage after fast interlayer transport g_{f1} into the layer is completed at $t = t_p$, thermal interlayer transport g_{t1} into the layer is measured at $t = t_d$. For detailed explanation see text.

in three groups based on their shape. The first group of steps that are flat occurs near 0.5 coverage and indicates that no net interlayer transport occurs into or out of the layer near half coverage. The second group of steps has a slightly downward curvature that indicates interlayer transport out of the layer at coverage below 0.5. The third group of steps has a slightly upward curvature that indicates interlayer transport into the growing layer above coverage 0.5. The curvature of the non-flat steps becomes more pronounced with increasing dwell time, indicating increasing interlayer transport with time between successive laser shots. The curvature also increases with increasing growth temperature. Based on the temperature dependence and the sluggish nature of the process during the dwell time we conclude that this component corresponds to thermally driven interlayer transport.

The coverage transients provide a quantitative picture of the closeness of S2L growth in STO to LBL growth. The defining feature of LBL growth is that the single growing layer must be full before a new layer can start growing. Although LBL is not realizable physically it represents a useful benchmark for comparison. Indeed, Fig. 9.8, to be discussed later, shows clearly that LBL growth is not happening in STO PLD growth. The dwell time also plays an important role in determining the quality of LBL growth. For example, Fig. 9.8 shows that when the base layer coverage reaches 0.9 the coverage on the top of a base layer is about 0.3 for a dwell time of 50 s and less than 0.2 for a dwell time of 0.2 s. These trends indicate that PLD growth becomes more like LBL growth as the dwell times become shorter. In connection with this trend, it will be shown below that the thermal interlayer transport component is not essential for PLD film growth.

9.3.4 The crystallization timescale

The crucial question for understanding the mechanism of PLD is what happens during the very fast interlayer transport step. The first approximation for the timescale of this step is the crystallization time. Another advantage of using a third generation synchrotron is that the high brightness permits measurement of the structural changes with the highest time resolution available. The crystallization time refers to the time when a crystalline structure formation can be detected on the growing surface. This initial crystallization represents an upper limit for the combination of surface transport steps that must occur to enable incorporation at the proper lattice site. This measurement is performed using a reflection that has a momentum transfer component parallel to the growing surface. In MBE growth of Ge it was observed that the intensity of the $(0\ 0\ \frac{1}{2})$ specular rod falls immediately after the shutter is opened (Vlieg et al., 1988). In contrast, there was a few seconds' delay before the intensity started falling in the off-specular reflection after the Ge shutter was opened. This delay is attributed to the time that it takes the first arriving Ge atoms to find the right crystallographic sites for incorporation into the lattice. The delay was found to decrease with increasing substrate temperature indicating the presence of thermally driven surface transport.

The crystallization time is also important for understanding the mechanism of PLD growth. Based on the timescale of crystallization it is possible to determine whether the conventional picture that deposition and growth are separate in PLD is correct. The assumption that these two stages are separable is also the basis for the interpretation of the recovery in some earlier PLD growth experiments using RHEED. The recovery in the RHEED intensity where only the slow thermally driven component was observed was attributed to crystallization of the plume species (Karl and Stritzker, 1992). In our experiments the time delay between the arrival of the plume and crystallization was determined by simultaneously measuring the intensity of the (0 0 ½) specular and the (0 1 ½) off-specular rods of STO. The absence of a measurable time delay between the fast transients for these two rods shown in Fig. 9.12 indicates that crystallization is instantaneous on the timescale of the measurement. The single shot transient data measured with time resolution in the μs range show that the instantaneous drop and jump occur on the same time scale and confirm that crystallization occurred faster than our measurements. It is known from independent measurements that plume arrival also occurs in a few μs (Wood et al., 1997; Sambri et al., 2008). Based on the similar timescales for plume arrival and crystallization we conclude that crystallization occurs during the arrival of the plume, suggesting that the interpretation of the slowest timescale to be crystallization is incorrect. The overlapping of the plume arrival timescale and the crystallization time scale also implies that separation of deposition and crystallization processes requires microsecond or faster pulse deposition timescales.

9.3.5 Quantitative determination of interlayer transport components

The simultaneous presence of two layers in S2L growth creates three exposed levels (see Fig. 9.10). The expected coverage f_n (n = 1–3) for these three surfaces is given by the exposed coverage ($\theta_n - \theta_{n+1}$) and the shot number p needed to deposit one layer. The expected coverage on top of the substrate is the fraction of the pulse that lands into the holes, $f_1 = (1 - \theta_1)/p$. This material stays in the holes and contributes to the growth of the base (first) layer, $\Delta\theta_1$. The growth of the layer $\Delta\theta_n$ (n = 1,2) is the coverage that is found as crystalline STO immediately following the laser shot. The expected coverage on top of the base layer (islands on the substrate) is given by $f_2 = (\theta_1 - \theta_2)/p$. Figures 9.10(b) and 9.10(c) show that f_2 consists of three components. The first component stays on top of the layer and contributes to the growth of the top (second) layer $\Delta\theta_2$. The second component corresponds to the amount

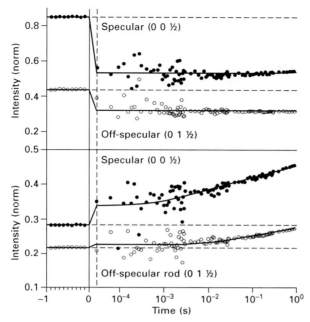

9.12 The measurement of fast transients for the specular (0 0 ½) and off-specular (0 1 ½) rods with μs scale resolution. Note that the data are plotted on a logarithmic scale that spans the time 1 s before and 1 s after the laser shot. The top shows a drop in the first shot following the maximum intensity, and the bottom shows a jump in the first shot following the minimum. The horizontal dashed lines mark the intensity before the laser shot. The vertical solid line marks $t = 0$ when the laser fires, the vertical dashed line corresponds to t_p, the end of the plume arrival time, and the end of the time axis corresponds to t_d, the dwell time.

transferred by the fast interlayer transport step into the base layer, g_{f_1}. And the third component shown in Fig. 9.10(c) is the amount transferred during the dwell time by the slow thermal step into the base layer, g_{t_1}. The expected coverage $f_3 = (\theta_2 - \theta_3)/p$ comes from the fact that the plume also lands on top of the second layer islands. However, as long as the second layer islands are smaller than λ (below coverage 0.6) the third layer cannot start growing and the entire amount must completely transfer down onto the base layer by fast interlayer transport, g_{f_2}. The fraction f_3 increases with the island size and just before coalescence it is more than half of the incoming plume. The quantitative determination of these interlayer transport fractions is enabled by the separation of the timescales for thermal and fast interlayer transport that is possible because of the four orders of magnitude difference in their magnitude.

The thermal interlayer transport component, g_{t_n} ($n = 1,2$) is the simplest to determine. This amount can be read off directly from the time-dependent coverage plots in Figs 9.8 and 9.11(a) and 9.11(b) as the difference between the coverages corresponding to the slow stage at the end (t_d) and the beginning of the dwell time (t_p). A positive value means that the material transfers into the layer and a negative value means that it transfers out of the layer. The fast interlayer transport component, g_{f_n} ($n = 1,2$) is also straightforward to determine. As already shown, the expected coverage f_n ($n = 1-3$) of the particular layer after the laser shot (θ^{m+1}) shown in Figs 9.11(c) and 9.11(d) at $t = 0$ is calculated from the known coverages before laser shot m, θ^m. The actual coverage determined immediately after the laser shot $t = t_p$ is different because fast interlayer transport already occurred before the measurement was performed. The difference between the expected coverage and the actual coverage (at the beginning and the end of the slanted segment in Figs 9.11(c) and 9.11(d) corresponds to the fast interlayer transport fraction for that particular laser shot.

After all the interlayer transport components are determined, the growth of the layer $\Delta\theta_n$ ($n = 1,2$) is computed depending on whether the layer is in the pre-coalescence stage or in the post-coalescence stage. In the pre-coalescence stage the net fast interlayer transport $\Delta g_{f_2} = g_{f_2} - g_{f_1}$ is first negative (out of the layer). With increasing coverage it turns positive (into the layer), and in the post coalescence stage fast interlayer transport $\Delta g_{f_1} = g_{f_1}$ always occurs into the layer. As would be expected, the growth of the layer mirrors the trends in fast interlayer transport. In the pre-coalescence stage the growth of the layer $\Delta\theta_2 = f_2 + g_{f_2} - g_{f_1}$ consists only of the material that is left behind after the fast interlayer transport out of the layer was completed. In the post-coalescence stage the growth of the layer $\Delta\theta_1 = f_1 + g_{f_1}$ has two components. The first component is the fraction deposited directly into the holes (f_1) and the second component is the material that transferred into the layer by fast interlayer transport. Note that the slow interlayer transport component does not figure in the growth of the layer ($\Delta\theta_n$) calculations because the fast crystallization implies that the growth is complete at the end of the plume arrival, t_p.

The entire PLD growth of STO is summarized by the plots in Fig. 9.13. Because the shape of the time-dependent coverages depends on the initial coverage, the magnitudes of the fast interlayer transport g_{f_n}, the growth of the layer $\Delta\theta_n$, and the thermal interlayer transport g_{t_n} that are derived from the time-dependent coverages will also be coverage-dependent. The first plot Fig. 9.13(b) shows the fast interlayer transport as a function of the number of laser shots. These coverage values are normalized such that it takes 10 shots ($p = 1/10$) to deposit 1 u.c. of STO. Negative values indicate transport out of the layer and positive values designate transport into the layer. Although, only the values for one growing layer are plotted, these plots are best understood

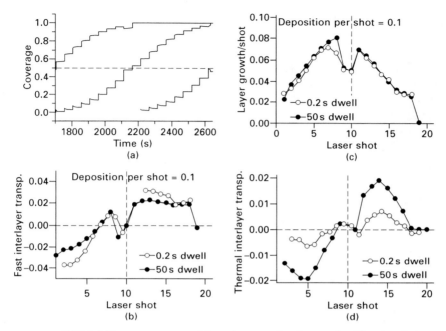

9.13 (a) The coverages of three successive layers for 50 s dwell time are replotted from Fig. 9.8 to aid understanding this figure. (b) Fast interlayer transport, (c) growth of the layer, and (d) thermal interlayer transport plotted as a function of the number of laser shots for 0.2 and 50 s dwell times. For detailed explanation see text.

by making a mental reference to the coverage of the second growing layer (see Fig. 9.8) for convenience replotted in Fig. 9.13(a), and remembering that complete filling of a layer occurs over time that corresponds to the deposition of two new layers. The time-dependent coverage data in Figs 9.8 and 9.13(a) show that the growing layer is always in communication with two other layers. Below 0.5 coverage the communication is with the layer below (base layer). Communication with the second (top) layer is established above 0.5 coverage when a new layer nucleates on top of the growing layer after the base layer is filled.

The coverage of the base layer just before coalescence is a convenient reference point for the zero of the coverage axis. At this coverage r is still smaller than λ, the length scale required for nucleation of a new layer, and the coverage of the top layer must be zero, meaning that complete interlayer transfer out of the layer (largest negative value) occurs. With the next shot coalescence is reached and new islands that nucleate on top reduce the amount of material available for interlayer transport out of the layer. With the following shots the islands are growing larger, slowing down further interlayer transport out of the layer. As the base layer approaches full coverage there are fewer holes remaining and interlayer transfer approaches zero. With completion of the base layer interlayer transport can occur only into the top layer and the value turns positive and increases with subsequent shots until coalescence is reached. After coalescence, interlayer transport into the layer slows down as the number of holes into which interlayer transport can occur decreases rapidly when approaching full coverage.

The second plot Fig. 9.13(c) shows that the growth-of-a-layer $\Delta\theta_n$ as a function of the number of laser shots mirrors nicely the fast interlayer transport plot. $\Delta\theta_2$ reaches a maximum when the base layer θ_1 is near filling up and the top layer coverage is ~0.4. Because effectively there is only one layer open here, there can be little or no interlayer transport and all material must go into $\Delta\theta_2$. However, because of the two layer nature of growth a laser shot going 100% in a single layer is never observed. Even at the maximum (near 0.4 coverage) less than 90% of the laser shot contributes to the growth of a single layer. Past the maximum, $\Delta\theta_2$ develops a downward trend because the increasing layer coverage θ_2 effectively blocks the surface and decreases the exposed coverage $(\theta_1 - \theta_2)$ and with that the fraction f_2 of the plume that is directly contributed to $\Delta\theta_2$.

The behavior of the slow thermally driven interlayer transport component shown in Fig. 9.13(d) is the easiest to understand. A distinctive feature of this plot is that it clearly depends on the dwell time. The amount of material transferred is determined from the data in Fig. 9.8 to increase with increasing dwell time from about 5% of a single shot for 0.2 s dwell time to about 20% of a single shot for 50 s dwell time. This behavior represents clear confirmation of the thermal nature of this interlayer transport component. The thermal interlayer transport component corresponds to transport of already crystallized STO unit cells. The plot in Fig. 9.13(d) shows that the transport out of the layer below 0.5 coverage is balanced by the transport into the layer above 0.5 coverage.

9.3.6 Diffuse scattering

Another advantage of using synchrotron X-rays is that they provide sufficient intensity to perform time-resolved measurements of the very weak diffuse scattering component that originates from incoherent fluctuations in local order and lateral organization (Fuoss et al., 1992; Robinson et al., 1996). Diffuse scattering measurements are performed by using an area detector, such as a CCD camera to capture the entire diffraction profile including both the sharp central peak and the broad diffuse scattering. Presented in the form of two-dimensional diffraction intensity maps shown in Fig. 9.14, diffuse scattering data provide information on the size of islands as illustrated by the change of the q (nm^{-1}) distribution of intensity as a function of deposition given in terms of the number of laser shots. The variations in the grayscale

Real-time studies of epitaxial film growth using SXRD 265

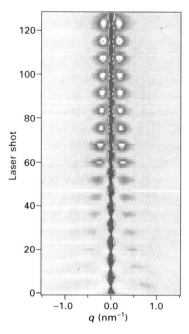

9.14 The complete diffraction intensity map from growth of 17 u.c. of STO obtained by measuring the (0 0 ½) diffraction profile using a CCD for growth at 10 s dwell time at 680 °C. The spine of the plot corresponds to the growth intensity oscillations of the specular beam. The side lobes at $q \neq 0$ on both sides give the diffuse scattering that is indicative of a well-developed correlation between the islands.

shades are used to aid visualization of the intensity changes. These two-dimensional diffraction maps contain the complete information about the formation and the evolution of the surface structure during STO film growth. The vertical line that cuts through the diffraction maps at $q = 0$ describes the growth intensity oscillations of the sharp central peak discussed previously. The larger elliptical shape features appearing at $q \neq 0$ on both sides of the central peak correspond to diffuse scattering.

The diffuse scattering describes the spatial distribution of the islands on the growing surface (Fuoss *et al.*, 1992; Alvarez *et al.*, 1998; Ferguson *et al.*, 2009). The presence of the diffuse side lobes indicates that a well-defined correlation exists between the islands. The q dependence of these features given by $\Delta q = 2\pi/L$ describes a characteristic length scale in terms of L, the correlation length that corresponds to the island separation ℓ in the pre-coalescence stage of the growth. Therefore, the shifting of q of the diffuse side lobes toward $q = 0$ correspond to increasing island sizes. From time-dependent measurements it is estimated that these changes are slow and are in the timescale of seconds.

The relationship between the central peak and the diffuse side lobes in layer filling during growth of the first three layers is illustrated in Fig. 9.15. As Fig. 9.14 shows the diffuse scattering signal is weak and distributed over a wide q range when the islands are very small at the onset of growth. Less noisy data are obtained by using δ-scans with a point detector. As expected, the plot shows that the intense central peak and the diffuse scattering are oscillating out of phase. The first appearance of the weak diffuse scattering component is observed at $q \cong 0.15$ after the first laser shot. The island size estimated from $\Delta q = 2\pi/r$ is $r = 5$ nm. An increasing island density after the second laser shot would be manifested as an increase in q. Simple layer filling at a fixed island density with increasing island sizes would result only in the change of the intensity of the diffuse side lobes located at a fixed q. At a full layer the diffuse scattering should vanish and nucleation of the next layer would occur for a q equal or close to the value for the first shot. Instead the plots in Figs 9.14 and 9.15 show that q decreases continuously with subsequent laser shots. This behavior is unexpected for simple layer filling

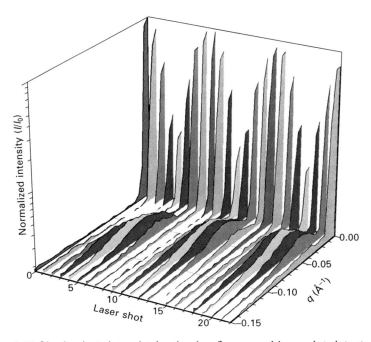

9.15 Single shot data obtained using δ-scans with a point detector showing the relative intensity of the specular and diffuse intensity peaks during growth of the first 3 u.c. of STO. The diffuse scattering oscillates out of phase with the specular intensity during layer filling.

and indicates that the island size is increasing during growth. This trend is characteristic of thermally driven processes such as island ripening. It was confirmed that q shifts to smaller values quicker for longer dwell times at a fixed temperature, and for increasing temperature at a fixed dwell time.

The shifting of q toward smaller values even when deposition of a new layer – referenced to the maximum intensity of the specular peak – is started, indicates the absence of independent renucleation on a flat growth surface. The dependence of the q for a newly growing layer on the q of the filling layer suggests that some type of communication exists between these two layers. This memory effect between the growth of two successive layers is a confirmation of the simultaneous growth of two layers in the S2L growth mode as concluded from the time-dependent coverage evolution. The most likely explanation for the existence of the memory effect is that interlayer transport makes the growth of two successive layers interdependent on each other. A careful examination of the scattering maps reveals that after the first few layers the diffuse scattering between two successive layers never fully vanishes in agreement with the conclusion that a flat growing surface is never again restored.

9.4　Future trends

The SXRD studies reveal new details of the PLD growth mechanism, leading to the conclusion that the potential of PLD is currently underutilized. The most important of these findings is that the actual film growth of STO in PLD – where growth is equivalent to initial crystallization confirmed by the first detectable presence of the crystalline STO phase – occurs on a much faster timescale than previously thought. The crystallization time in PLD as the upper limit for combination of surface transport needed for the arrival of the growth species at the proper crystallographic sites was determined to occur faster than it can be measured with μs scale time resolution. Using diffuse scattering measurements it was shown that a second much slower step that corresponds to continuous island ripening occurs throughout the growth process. The island ripening process occurs by thermally driven conversion of already crystalline smaller STO islands into larger ones. The formation of larger islands (by long dwell times) is not only unnecessary for PLD, it is actually detrimental to LBL growth because it reduces the nucleation density and favors premature nucleation of the second layer and eventually acceleration of the interface broadening and 3D growth. Longer dwell times and higher substrate temperatures increase the impact of thermal processes and diminish the non-equilibrium effects of PLD growth.

The S2L growth mode will persist indefinitely during high quality growth and represents the highest perfection growth mode physically possible by any growth method. Post-growth AFM characterizations show that STO films grown in the S2L growth mode reveal no interface broadening even after the growth of more than 100 u.c. thick STO. In terms of a kinetic picture it is now clear why PLD is so similar to LBL growth. Although the presence of the second layer certainly interferes and slows down fast interlayer transport into the base layer, it appears that interlayer transport is sufficiently effective to enable a 'stretched out' LBL growth mode to complete one full layer over two periods.

It is appealing to think about a 'pure' PLD growth process in which the thermal component is suppressed and the surface smoothing effects of energetic deposition (non-equilibrium growth) from the plume are fully preserved. The first step toward realizing such a process is to reduce the dwell time. An added benefit of eliminating or reducing the dwell times is an increase in the average growth rate that is presently used with PLD growth conditions not much higher than typical MBE growth rates of about 1 ML/min. It is concluded that an effective growth manipulation scheme for PLD film growth (Koster *et al.*, 1999) must be based on maximizing the role of the fast interlayer transport step and suppressing the contribution from the sluggish thermal interlayer transport component during the dwell time, by reducing the substrate temperature, or the dwell time, or a combination of the two.

The growth kinetics picture of STO PLD that emerges from SXRD data is part of a general framework that is also applicable in homo- and heteroepitaxial growth of other perovskites. The present studies of kinetics from single laser shot deposition show that departure from LBL growth in homoepitaxy happens as a consequence of impeded fast interlayer transport. This is the factor that is most likely to adversely affect LBL growth in heteroepitaxy as well. But the mechanisms through which the interactions between the dissimilar atoms of the substrate and the film can alter the rate of interlayer transport in layer filling at the onset of growth are not known. The macroscopic manifestation of these interactions includes 3D growth, strain, chemical inhomogeneity, and phase change that fundamentally alter the properties of the near-interface region. Manipulating the growth kinetics becomes an indispensible tool in heteroepitaxial growth because it provides a way to open film growth pathways that are closed by thermodynamics in near-equilibrium growth methods. The special importance that the growth of the first layer plays in homoepitaxial kinetics becomes magnified in heteroepitaxy by the fact that it is both kinetically and structurally different from the rest of the layers that make up the film. Direct determination of interlayer transport rates from the time-dependent coverages, island size distributions, and island size evolution from diffuse scattering can be used for understanding how interface sharpness will be affected by island nucleation and island growth governed by the additional factors that are present in heteroepitaxial growth.

9.5 Acknowledgment

Research sponsored by the Materials Sciences and Engineering Division, Basic Energy Sciences (BES), US Department of Energy (DOE), and performed in part at the Advanced Photon Source, a DOE-BES user facility.

9.6 References

Adamovic, D., Chirita, V., Munger, E.P., Hultman, L. & Greene, J.E. (2007), Kinetic pathways leading to layer-by-layer growth from hyperthermal atoms: a multibillion time step molecular dynamics study, *Physical Review B*, **76**, 115418–115425.

Als-Nielsen, J. & McMorrow, D. (2001), *Elements of modern x-ray physic*, New York, John Wiley.

Alvarez, J., Lundgren, E., Torrelles, X. & Ferrer S. (1998), Determination of scaling exponents in Ag(100) homoepitaxy with X-ray diffraction profiles, *Physical Review B*, **57**, 6325–6328.

Aziz, M.J. (2008), Film growth mechanisms in pulsed laser deposition, *Applied Physics A*, **93**, 579–587.

Blank, D.H.A., Rijnders, G.J.H.M., Koster, G. & Rogala, H. (1999), *In situ* monitoring by reflective high-energy electron diffraction during pulsed-laser deposition, *Applied Surface Science*, **138–139**, 17–23.

Braun, W. (1999), *Applied RHEED: reflection high-energy electron diffraction during crystal growth*, Berlin, Heidelberg, Springer-Verlag.

Braun, W. & Ploog, K.H. (2006), *In situ* studies of epitaxial growth by synchrotron X-ray diffraction, *Surface Review and Letters*, **13**, 155–166.

Braun, W., Daweritz, L. & Ploog, K.H. (1998), Origin of electron diffraction oscillations during crystal growth, *Physical Review Letters*, **80**, 4935–4938.

Charlton, G., Brennan, S., Muryn, C.A., McGrath, R. & Norman, D. (2000), Surface relaxation of $SrTiO_3$ (001), *Surface Science Letters*, **457**, L376.

Chern, M.Y., Gupta, A., Hussey, B.W. & Shaw, T.M. (1993), Reflection high-energy electron diffraction intensity monitored homoepitaxial growth of $SrTiO_3$ buffer layers by pulsed-laser deposition, *Journal of Vacuum Science and Technology A* **11**, 637–641.

Christen, H.M. & Eres, G. (2008), Recent advances in pulsed-laser deposition of complex oxides, *Journal of Physics: Condensed Matter*, **20**, 264005.

Clarke, S. & Vvedensky, D.D. (1987), Origin of reflection high-energy electron-diffraction intensity oscillations during molecular-beam epitaxy: a computational modeling approach, *Physical Review Letters*, **80**, 2235–2238.

Cohen, P.I., Petrich, G.S., Pukite, P.R., Whaley, G.J. & Arrott, A.S. (1989), Birth–death models of epitaxy: I. Diffraction oscillations from low index surfaces, *Surface Science*, **216**, 222–248.

Dagotto, E. (2007), When oxides meet face to face, *Science*, **318**, 1076–1077.

Doolittle, W.A., Carver, A.G. & Henderson W. (2005), Molecular beam epitaxy of complex metal-oxides: where have we come, where are we going, and how are we going to get there? *Journal of Vacuum Science and Technology B*, **23**, 1272–1276.

Ehrlich, G. & Hudda, F.G. (1966), Atomic view of surface self-diffusion: tungsten on tungsten, *Journal of Chemical Physics*, **44**, 1939.

Eres, G., Tischler, J., Yoon, M., Larson, B., Rouleau, C., Lowndes, D. & Zschack, P.

(2002), Time-resolved study of SrTiO$_3$ homoepitaxial pulsed-laser deposition using surface X-ray diffraction, *Applied Physics Letters*, **80**, 3379–3381.

Evans, J.W., Thiel, P.A. & Bartelt, M.C. (2006), Morphological evolution during epitaxial thin film growth: formation of 2D islands and 3D mounds, *Surface Science Reports*, **61**, 1–128.

Evans-Lutterodt, K.W. & Tang, M.T. (1995), Angle calculations for a '2+2' surface X-ray diffractometer, *Journal of Applied Crystallography*, **28**, 318.

Farrow, R.F.C. (1995), *Molecular beam epitaxy: Applications to key materials*, Park Ridge, Noyes Publications.

Feidenhans'l, R. (1989), Surface structure determination by X-ray diffraction, *Surface Science Reports*, **10**, 105–188.

Ferguson, J.D., Arikan, G., Dale, D.S., Woll, A.R. & Brock, J.D. (2009), Measurements of surface diffusivity and coarsening during pulsed laser deposition, *Physical Review Letters*, **103**, 256103.

Fleet, A., Dale, D., Suzuki, Y. & Brock, J.D. (2005), Observed effects of a changing step-edge density on thin-film growth dynamics, *Physical Review Letters*, **94**, 036102.

Fuoss, P.H. & Brennan, S. (1990), Surface sensitive X-ray scattering, *Annual Review of Materials Science*, **20**, 365–390.

Fuoss, P.H., Kisker, D.W., Lamelas, F.J., Stephenson, G.B., Imperatori, P. & Brennan, S. (1992), Time-resolved X-ray scattering studies of layer-by-layer epitaxial growth, *Physical Review Letters*, **69**, 2791–2794.

Giacovazzo, G., Monaco, H.L., Viterbo, F., Gilli, G., Zanotti, G. & Catti, M. (2011), *Fundamentals of crystallography*, Oxford, Oxford University Press.

Gorbenko, O.Y., Samoilenkov, S.V., Graboy, I.E. & Kaul, A.R. (2002), Epitaxial stabilization of oxides in thin films, *Chemistry of Materials*, **14**, 4026–4043.

Herger, R., Willmott, P.R., Bunk, O., Schleputz, C.M., Patterson, B.D. & Delley, B. (2007), Surface of strontium titanate, *Physical Review Letters*, **98**, 076102.

Hinnemann, B., Hinrichsen, H. & Wolf, D.E. (2001), Unusual scaling for pulsed laser deposition, *Physical Review Letters*, **87**, 135701.

Huijben, M., Brinkman, A., Koster, G., Rijnders, G., Hilgenkamp, H. & Blank, D.H.A. (2009), Structure–property relation on SrTiO$_3$/LaAlO$_3$ interfaces, *Advanced Materials*, **21**, 1665–1677.

Ichimiya, A. & Cohen, P.I. (2004), *Reflection high-energy electron diffraction*, Cambridge, Cambrige University Press.

Ingle, N.J.C., Yuskauskas, A., Wicks, R., Paul, M. & Leung, S. (2010), The structural analysis possibilities of reflection high energy electron diffraction, *Journal of Physics D: Applied Physics*, **43**, 133001.

Jacobsen, J., Cooper, B.H. & Sethna, J.P. (1998), Simulations of energetic beam deposition: from picoseconds to seconds, *Physical Review B*, **58**, 15847–15865.

Jang, H.W., Felker, D.A., Bark, C.W., Wang, Y., Niranjan, M.K., Nelson, C.T., Zhang, Y., Su, D., Folkman, C.M., Baek, S.H., Lee, S., Janicka, K., Zhu, Y., Pan, X.Q., Fong, D.D., Tsymbal, E.Y., Rzchowski, M.S. & Eom, C.B. (2011), Metallic and insulating oxide interfaces controlled by electronic correlations, *Science*, **331**, 886–889.

Joyce, B.A. & Joyce, T.B. (2004), Basic studies of molecular beam epitaxy – past, present and some future directions, *Journal of Crystal Growth*, **264**, 605–619.

Kaganer, V.M. (2007), Crystal truncation rods in kinematical and dynamical X-ray diffraction theories, *Physical Review B*, **75**, 245425.

Karl, H. & Stritzker, B. (1992), Reflection high-energy electron diffraction oscillations modulated by laser-pulse-deposited YBa$_2$Cu$_3$O$_{7-x}$, *Physical Review Letters*, **69**, 2939–2942.

Kawasaki, M., Takahashi, K., Maeda, T., Tsuchiya, R., Shinohara, M., Ishiyama, O., Yonezawa, T., Yoshimoto, M. & Koinuma, H. (1994), Atomic control of the $SrTiO_3$ crystal surface, *Science*, **266**, 1540.

Korte, U. & Maksym, P.A. (1997), Role of the step density in reflection high-energy electron diffraction: questioning the step density model, *Physical Review Letters*, **78**, 2381–2384.

Koster, G., Kropman, B.L., Rijnders, G.J.H.M., Blank, D.H.A. & Rogalla H. (1998), Quasi-ideal strontium titanate crystal surfaces through formation of strontium hydroxide, *Applied Physics Letters*, **73**, 2920–2922.

Koster, G., Rijnders, G.J.H.M., Blank, D.H.A. & Rogalla H. (1999), Imposed layer-by-layer growth by pulsed laser interval deposition, *Applied Physics Letters*, **74**, 3729–3731.

Lam, P.-M., Liu, S.J. & Woo, C.H. (2002), Monte Carlo simulation of pulsed laser deposition, *Physical Review B*, **66**, 045408.

Larsson, M.I., Ni, W.X. & Hansson, G.V. (1995), Manipulation of nucleation by growth rate modulation, *Journal of Applied Physics*, **78**, 3792.

Lee, H.N., Christen, H.M., Chisholm, M.F., Rouleau, C.M. & Lowndes, D.H. (2005), Strong polarization enhancement in asymmetric three-component ferroelectric superlattices, *Nature*, **433**, 395–399.

Lippmaa, M., Nakagawa, N., Kawasaki, M., Ohasi, S. & Koinuma, H. (2000), Growth mode mapping of $SrTiO_3$ epitaxy, *Applied Physics Letters*, **76**, 2439–2441.

Lowndes, D.H., Geohegan, D.B., Puretzky, A.A., Norton, D.P. & Rouleau, C.M. (1996), Synthesis of novel thin-film materials by pulsed laser deposition, *Science*, **273**, 898–903.

Mannhart, J. & Schlom, D.G. (2010), Oxide interfaces – an opportunity for electronics, *Science*, **327**, 1607–1611.

Michely, T. & Krug, J. (2004), *Islands, mounds and atoms*, Springer Series in Surface Sciences vol 42, Berlin, Springer-Verlag.

Ohnishi, T., Shibuya, K., Lippmaa, M., Kobayashi, D., Kumigashira H., Oshima, M. & Koinuma, H. (2004), Preparation of thermally stable TiO_2-terminated $SrTiO_3$(100) substrate surfaces, *Applied Physics Letters*, **85**, 272–274.

Ohtomo, A. & Hwang, H.Y. (2004), A high-mobility electron gas at the $LaAlO_3/SrTiO_3$ heterointerface, *Nature*, **427**, 423–426.

Pimpinelli, A. & Villan, J. (1998), *Physics of crystal growth*, Cambridge, Cambridge University Press.

Reiner, J.W., Walker, F.J. & Ahn, C.H. (2009), Atomically engineered oxide interfaces, *Science*, **323**, 1018–1019.

Rijnders, G. & Blank, D.H.A. (2005), Real-time monitoring by high-pressure RHEED during pulsed laser deposition, in *Thin Films and Heterostructures for Oxide Electronics*, ed. S.B. Ogale, New York, Springer.

Robinson, I.K. (1986), Crystal truncation rods and surface roughness, *Physical Review B*, **33**, 3830–3836.

Robinson, I.K. & Tweet, D.J. (1992), Surface X-ray-diffraction, *Reports on Progress in Physics*, **55**, 599–651.

Robinson, I.K., Whiteaker, K.L. & Walko, D.A. (1996), Cu island growth on Cu(110), *Physica B*, **221**, 70–76.

Rosenfeld, G., Poelsema, B. & Comsa, G. (1997), Epitaxial growth modes far from equilibrium, p. 66 in *Growth and properties of ultrathin epitaxial layers*, The chemical physics of solid surfaces, Vol. 8, ed. by King, D.A. & Woodruff, D.P., Elsevier, Amsterdam

Saldin, D.K. & Shneerson, V.L. (2008), Direct methods for surface crystallography, *Journal of Physics: Condensed Matter*, **20**, 304208.

Sambri, A., Amoruso, A., Wang, X., Miletto Granozio, F. & Bruzzese, R. (2008), Plume propagation dynamics of complex oxides in oxygen, *Journal of Applied Physics*, **104**, 053304.

Schlom, D.G., Haeni, J.H., Lettieri, J., Theis, C.D., Tian, W., Jiang, J.C. & Pan, X.Q. (2001), Oxide nano-engineering using MBE, *Materials Science and Engineering B*, **87**, 282–291.

Schlom, D.G., Chen, L.-Q., Pan, X., Schmehl, A. & Zurbuchen, M.A. (2008), A thin film approach to engineering functionality into oxides, *Journal of American Ceramic Society*, **91**, 2429–2454.

Schmid, M., Lenauer, C., Buchsbaum, A., Wimmer, F., Rauchbauer, G., Scheiber, P., Betz, G. & Varga, P. (2009), High island density in pulsed laser deposition: causes and implications, *Physical Review Letters*, **103**, 076101.

Schwoebel, R.L. & Shipsey, E.J. (1966), Step motion on crystal surfaces, *Journal of Applied Physics*, **37**, 3682.

Shin, B. & Aziz, M.J. (2007), Kinetic-energy induced smoothening and delay of epitaxial breakdown in pulsed-laser deposition, *Physical Review B*, **76**, 085431.

Sutton, A.P. & Balluffi, W. (1995), *Interfaces in crystalline materials*, Oxford, Clarendon Press.

Taylor, M.E. & Atwater, H.A. (1998), Monte Carlo simulations of epitaxial growth: comparison of pulsed laser deposition and molecular beam epitaxy, *Applied Surface Science*, **127–129**, 159.

Tersoff, J., Dernier van der Gon, A.W. & Tromp, R.M. (1994), Critical island size for layer-by-layer growth, *Physical Review Letters*, **72**, 266–269.

Tischler, J.Z., Eres, G., Larson, B.C., Rouleau, C.M., Zschack, P. & Lowndes, D.H. (2006), Nonequilibrium interlayer transport in pulsed laser deposition, *Physical Review Letters*, **96**, 226104.

Tromp, R.M. & Hannon, J.B. (2002), Thermodynamics of nucleation and growth, *Surface Review and Letters*, **9**, 1565–1593.

van der Vegt, H.A., van Pinxteren, H.M., Lohmeier, M., Vlieg, E. & Thornton, J.M.C. (1992), Surfactant induced layer-by-layer growth of Ag on Ag (111), *Physical Review Letters*, **68**, 3335–3338.

Vasco, E. & Sacedón, J.L. (2007), Role of cluster transient mobility in pulsed laser deposition-type growth kinetics, *Physical Review Letters*, **98**, 036104.

Vlieg, E., Dernier van der Gon, A.W., van der Veen, J.F., Macdonald, J.E. & Norris, C. (1988), Surface X-ray scattering during crystal growth: Ge on Ge(111), *Physical Review Letters*, **61**, 2241–2244.

Vlieg, E., Lohmeier, M. & van der Vegt, H.A. (1995), Surface X-ray crystallography of growing crystals and interfaces, *Nuclear Instruments and Methods Research B*, **97**, 358–363.

Warrender, J.M. & Aziz, M.J. (2007), Kinetic energy effects on morphology evolution during pulsed laser deposition of metal-on-insulator films, *Physical Review B*, **75**, 085433.

Willmott, P.R. (2004), Deposition of complex multielemental thin films, *Progress in Surface Science*, **76**, 163–217.

Willmott, P.R. & Huber, J.R. (2000), Pulsed laser vaporization and deposition, *Reviews of Modern Physics*, **72**, 315–328.

Willmott, P.R., Herger, R., Schlepütz, C.M., Martoccia, D. & Patterson, B.D. (2006),

Energetic surface smoothing of complex metal-oxide thin films, *Physical Review Letters*, **96**, 176102.

Wood, R.F., Leboeuf, J.N., Puretzky, A.A. & Geohegan, D.B. (1997), Dynamics of plume propagation and splitting during pulsed-laser ablation, *Physical Review Letters*, **79**, 1571–1574.

Woodruff, D.P. & Delchar, T.A. (1994), *Modern techniques of surface-science*, Cambridge, Cambridge University Press.

Zangwill, A. (1988), *Physics at surfaces*, Cambridge, New York, Cambridge University Press.

Zhang, Z. & Lagally, M. (1997), Atomistic processes in the early stages of thin-film growth, *Science*, **276**, 377.

Zinke-Allmang, M. (1999), Phase separation on solid surface: nucleation, coarsening and coalescence kinetics, *Thin Solid Films*, **346**, 1.

Zubko, P., Gariglio, S., Gabay, M., Ghosez, P. & Triscone, J.-M. (2011), Interface physics in complex oxide heterostructures, *Annual Review of Condensed Matter Physics*, **2**, 141–165.

Index

ambient pressure XPS, 89–90
angle-resolved photoemission spectroscopy, 57–61
   crystal energy levels in initial state and measured photoemission spectrum, 58
   illustration, optical (direct) transitions from electrons in crystal band and ARPES, 59
   three-step model illustration, 61
anti-Bragg point, 246
atomic force microscopy, 111, 199
atomic layer deposition, 107, 156, 163
Auger electron spectroscopy, 189, 201–6
   chemical information from the Auger line shape and energy position, 205–6
   experimental set-up, 201–2
   growth control of lanthanide alloys, 204–5
      dysprosium on a Si wafer, 204
   growth monitoring of MgO and of $Cr_xMo_{(1-x)}$, 202, 204
   IMFP data, 201
   oxygen monitoring during ZnO oxide growth, 202
      ZnO growth on GaN substrate, 203
   surface purity control, 206
      phosphor impurity Auger line, 207

backscatter Kikuchi diffraction
   dual-screen RHEED and Kikuchi pattern collection, 37–40
      astrodome geometry and dual-screen geometry, 38
      Kikuchi pattern of $SrRuO_3$ film grown on $SrTiO_3$ substrate, 40
   epitaxial film strain determination, 41–2
      BKP circular subsection from Si after contrast enhancement, 42
   epitaxial film structure determination, 45–8
      high-pressure phase of $SrCuO_2$ with and without buckled Cu–O planes, 47
      Kikuchi lines from CuFeS intersecting within Kikuchi band, 46
   inelastic scattering techniques, film growth *in situ* characterization, 29–48
      dual-screen RHEED and Kikuchi pattern collection, 37–40
      epitaxial film strain determination, 41–2
      epitaxial film structure determination, 45–8
      Kikuchi lines in RHEED images, 33–7
      Kikuchi patterns, 30–3
      kinematic and dynamic scattering, 42–5
      lattice parameter determination, 40–1
   Kikuchi lines in RHEED images, 33–7
      crystal surface geometry, 34
      RHEED and Kikuchi pattern, 37
   kinematic and dynamic scattering, 42–5
      BKP of GaN at 20 kV, experimental pattern and dynamical simulation, 44

backscatter Kikuchi patterns, 31
backscattered electrons, 182–3
Balzers QMS, 220
band transitions, 187
Bayard–Alpert configuration, 219
beam wobble
  sample rotation, 124–6
    aligned to rotation axis, 125
    return-path ellipsometry method, 126
Beer's law, 223, 225
biological films, 137–41
  600 nm wavelength and SE determined thickness during *in situ* liquid-cell experiment, 139
  SE and QCM thickness measured during CTAB adsorption and rinse process, 141
blocking dips, 158
Bohr electron velocity, 159
Boltzmann's constant, 57
Bragg condition, 9
Bragg diffraction, 30–1
BST, 156

CASINO, 183–4
cathodoluminescence spectroscopy, 192–3
  experimental set-up, 192–3
  measurement of substrate temperature, 193
cetyltrimethylammonium bromide, 138
characteristic energy losses, 184–8, 205
  continuous X-ray emission, 187–8
  inner shell ionization processes, 184–5
  ionization cross sections, 185
  plasmons and band transitions, 186–7
    probability for multiple energy loss, 187
chopped ion gauge, 218–20
collimators, 232
colossal magnetoresistance, 69
common pseudo-substrate approximation, 111–13
complementary metal oxide semiconductor, 156
compound semiconductors, 132–3
  real-time feedback control demonstration, 134
  SE measurements vs *ex situ* Fourier transform infrared transmission results, 133
convergent beam electron diffraction, 31
Coster–Kronig transitions, 188
crystal truncation rods, 247
crystallization time, 259–60
  fast transients measurement, 261

derivative spectroscopy *see* wavelength-modulation spectroscopy
diamond-like carbon (DLC) films, 130
diffraction theory, 18
diffuse scattering, 264–7
  complete diffraction intensity map, 265
  single shot data for relative intensity of the specular and diffuse intensity peaks, 266
direct recoil spectroscopy, 159–61

effective attenuation length (EAL), 77
effective medium approximation, 119
Ehrlich–Schwoebel barrier, 242, 254
elastic mean free path, 183–4
elastic recoil detection, 159
elastic scattering, 183–4
electron backscatter diffraction, 32
electron channelling patterns, 32
electron impact emission spectrometry, 222–3
electrostatic energy analyzer, 161–3
ellipsometry
  principles, 100–9
    basic elllipsometer components, 105
    data analysis, 103–6
    data analysis flowchart, 104
    elliptical polarization, 101
    in situ SE, 106–9
    in situ SE measurements during transparent layer growth, 108
    light beam interacts with thin film structure, 102
    optical constants for $Si_{1-x}Ge_x$ with different compositions, 106
    quartz crystal microbalance with dissipation monitoring (QCM-D) with liquid flow-cell and optical windows, 108

SE measurements from two transparent thin films on silicon substrate, 103
spectroscopic ellipsometer mounted to atomic layer deposition (ALD), 107
spectroscopic ellipsometry, 100–3
EMA theory, 134
epitaxial film growth
surface X-ray diffraction, 239–68
future trends, 267–8
growth intensity oscillations, 240–2
growth kinetics studies of pulsed laser deposition, 245–9
interlayer transport models, 242–5
Ewald sphere, 8

Fermi energy, 57
film growth
in situ characterization using backscatter Kikuchi diffraction, 29–48
dual-screen RHEED and Kikuchi pattern collection, 37–40
epitaxial film strain determination, 41–2
epitaxial film structure determination, 45–8
Kikuchi lines in RHEED images, 33–7
Kikuchi patterns, 30–3
kinematic and dynamic scattering, 42–5
lattice parameter determination, 40–1
Fourier transform, 9–10
Fourier transform infrared (FTIR) spectroscopy, 133
Frank–van der Merwe, 12
Fresnel equation, 105

GaAs, 170–2
general virtual interface, 113
growth intensity oscillations, 240–2
intensities and coverages for LBL growth, 242
growth rate
and etch rate, 109–11
dielectric layer film thickness, during sputter deposition, 110

film thickness and etch rate for PMMA thin film during plasma and etch process, 110
in situ SE measurements of AlGaAs film deposition, 113
optical constants, thin metal films from combined *in situ* SE and transmission intensity measurements, 112

Hiden QMS, 221
high-kinetic energy XPS, 92–4
experimental parameters and Si 1s XPS spectra, 94
Ni/Cu interface and Cu 2p 3/2 spectra taken at incident photon energy of 2010 eV, 95
high-order Laue zone HOLZ, 36
high-resolution X-ray diffraction, 41
higher harmonic generation, 63
hollow cathode lamps, 224

in situ spectroscopic ellipsometry characterization, 109–19
growth and etch rate, 109–11
multilayer and graded film, 114
process control, 114–15
process-dependent optical constant libraries, 116–18
surface and interface quality, 118–19
virtual interface approaches, 111–13
examples, 132–43
biological films, 137–41
compound semiconductors, 132–3
nanomaterials, 133–5
optical coatings, 135–7
photovoltaics, 142–3
in situ considerations, 119–31
mechanical integration, 119–24
offset light beam, 128
process conditions, 128–9
sample rotation and beam wobble, 124–6
software integration, 129–31
window concerns, 127–8
thin film growth characterization, 99–144
ellipsometry principles, 100–9

in situ considerations, 119–31
in situ examples, 132–43
in situ spectroscopic ellipsometry (SE) characterization, 109–19
inelastic mean free path, 75
inelastic scattering techniques
film growth in situ characterization using backscatter Kikuchi diffraction, 29–48
dual-screen RHEED and Kikuchi pattern collection, 37–40
epitaxial film strain determination, 41–2
epitaxial film structure determination, 45–8
Kikuchi lines in RHEED images, 33–7
Kikuchi patterns, 30–3
kinematic and dynamic scattering, 42–5
lattice parameter determination, 40–1
interlayer transport, 242–5
finite interlayer transport rate, 244
quantitative determination, 260–4
PLD growth of STO, 263
ion beam analysis
characterization of thin multicomponent films, 155–76
experimental set-ups, 161–5
schematic of the TOF–ion scattering and recoil spectroscopy system, 164
toroidal ion energy analyzer, 162
ion backscattering spectrometry and TOF ion scattering and recoil methods, 157–61
energy dependence of the stopping power, 160
schematics of binary collision process, 157
schematics of blocking and formation of the blocking dip in angular spectrum, 159
theory of IBA methods, 157–61
studies of film growth processes, 165–75
angular resolved ISS spectra of the SBT film, 175

areal density of Al and O, 172
growth mechanism model of an ALD HfO_2 layer, 171
Hf coverage, 170
in situ, real-time ISS spectra, 174
MEIS backscattering spectra, 167
MEIS spectra for GaAs sample, 172
MEIS spectra of Hf atoms, 169
MEIS spectra of Si atoms, 169
MSRI spectra, 166
MSRI spectra during SBT deposition, 173
O, Si and Ti diffusion processes, 168
ion scattering spectroscopy, 173–4

Kikuchi patterns, 30–3
analysis, 32–3
BKP from cleaved crystal of silicon, 33
Kikuchi band formation, 30
SEM, 32
TEM, 31
known orientation, 39

Laue circles, 8
Laue groups, 32

mass spectroscopy of recoiled ions, 165, 173–5
mean free path, 181, 189
mean squared error, 105
mechanical integration, 119–24
in situ SE to Kurt J. Lesker sputter deposition chamber, 121
reflected beam position at receiver plane, 123
with prism-assisted access to vacuum chamber through chamber lid, 122
medium energy ion scattering, 157, 163, 168, 171
metal-organic vapour phase epitaxy, 115
metal-organic chemical vapour deposition, 116
microchannel plates, 62
'mode-lock technology,' 216
molecular beam epitaxy, 12, 121, 181, 240
growth of oxide films with AA rate monitoring, 231–2

278   Index

Monte Carlo algorithm, 15
Monte Carlo simulations, 23–4
multi-beam dynamic model, 48
multicomponent oxides
  film growth processes, 165–75
  ferroelectric perovskite oxide thin film growth, 173–5
  in situ analysis of Hf oxide growth, 168–70
  initial stages of Al oxide film growth on GaAs, 170–2
multilayer film
  and graded film characterization, 114
  refraction profile index *versus* film thickness, 114

nanomaterials, 133–5
  film thickness measured by *in situ* SE during ALD growth of TiN and TaN films, 136
  growth rate determined from *in situ* SE plasma-assisted ALD, 136
  in situ SE measurements during silver nanoparticles curing in PVOH host, 135
nonvolatile ferroelectric random access memories, 165
nucleation distance, 252

one-step model, 60
optical absorption spectroscopy techniques, 223–9
  (HCL)-based AA vapour monitors, 225–7
    dual-beam HCL-AA system, 226, 227
    schematic of the MBE system, 225
    hollow cathode lamp vs tunable-diode laser source, 224
  tunable diode laser absorption spectroscopy, 228–9
    Doppler shift measurement, 229
    frequency doubled-diode-laser-based atomic absorption monitor, 228
optical coatings, 135–7
  refractive index calibration for PECVD deposited thin films mixtures of TiO_2 and SiO_2, 137

photoemission spectroscopy, 56

photomultiplier tube, 222
photovoltaic, 142–3
  optical constants for amorphous and nanocrystalline germanium films compared to crystalline germanium substrate, 142
planar optical model, 105
plasma-enhanced chemical vapour deposition, 113
poly(methacrylate), 110
poly(vinyl alcohol), 134
primary electrons
  processes and excitations, 181–8
prism-assisted method, 122
process condition, 128–9
  three liquid cell geometries, 130
process control, 114–15
  MBE growth control demonstration, 115
process-dependent optical constant libraries, 116–18
  in situ SE data from sputter process, 117
  optical constant library, 117
  refractive index at 633 nm and deposition rate *versus* bias voltage for aluminium oxide thin films, 116
  Ta and a-Si growth rate calibration *versus* gun current, 118
pulsed laser deposition, 3–5
  growth kinetics studies, 245–9
    experimental set-up, 247–9
    illustration of the (2+2)-circle diffractometer angles, 248
    intensity of crystal truncation rod, 247
    portable PLD chamber, 249
    use of SXRD, 246–7
    X-ray usage, 245–6
PZT, 165

quadrupole mass spectroscopy, 220–2
quartz crystal microbalance, 214–17
  dissipation monitoring, 107
  schematic of a quartz crystal monitor, 215
quartz crystal monitors, 140
quasi-elastic scattering, 43

reciprocity theorem, 32

reflection energy loss spectroscopy, 197–201
  example of REELS distributions, 198–9
  REELS of $SrTiO_3$ near the elastic reflected peak, 200
  experimental set-up, 198
  energy filtered RHEED diagrams, 199
  quantitative analyses of ionization loss, 200–1
  surface growth and roughness monitoring, 199–200
reflection high-energy electron diffraction (RHEED), 121–2
  and pulsed laser deposition (PLD), 3–5
    pulsed laser deposition chamber equipped with high-pressure RHEED, 4
  crystal growth: kinetics vs thermodynamics, 12–13
  in situ spectroscopies, 191–206
    Auger electron spectroscopy, 201–6
    CL spectroscopy, 192–3
    reflection energy loss spectroscopy, 197–201
    RHEED-TRAXS, 195–7
    XES combined with RHEED, 193–5
  in thin film growth in situ characterization, 3–26
  intensity variations and Monte Carlo simulations, 23–4
    simulated intensity variations during continuous deposition, 24
  kinetical growth modes and intensity response, 18–22
    kinetic growth modes during PLD, 19
    pulsed laser interval deposition, 21–2
    specular reflection intensity variations, 22
  patterns analysis: surface disorder, 8–12
    examples, 10–12
    geometrical information, 8–10
  principles, 5–8
    experiment set-up, 5
    patterns for different surface morphologies, 7
    solid electrons scattering, 6–8
  processes and excitations by primary electrons, 181–8
    elastic scattering and elastic mean free path, 183–4
    inelastic scattering processes and characteristic energy losses, 184–8
    linear range and mean free path for silicon, 183
    normal and grazing incidence angle trajectories of fast primary electrons, 182
    range and penetration depth, 184
  real-time in situ surface monitoring of thin film growth, 180–209
    comparison and future trends, 206–9
  recombination and emission processes, 188–91
    fluorescence yield for different shells as a function of the atomic number Z, 189
    quantification of signal intensities, 189–91
  specular intensity variation during deposition, 13–18
regression analysis, 105
residual resistivity ratio, 65
return-path ellipsometry method, 126
Rutherford backscattering spectrometry, 157, 165, 168, 199

Sauerbrey equation, 216
SBT, 156, 165, 173
scanning electron microscopy, 32
scanning probe microscopy, 5
Schrödinger equation, 8
Scienta analyser, 62
signal intensities, 189–91
  calibration model of atomic concentration, 190
simultaneous two-layer growth, 254–9
  AFM image of STO film surface, 256
  expected coverages for the three levels, 257
  pre-coalescence and post-coalesce time-dependent coverages, 258
  time-dependent coverages, 255
Snell's law, 103
software integration, 129–31

roll-to-roll coater incorporating multiple in-line SE systems, 131
top view, dual magnetron PECVD sputter chamber with planetary rotation, 131
software synchronization, 130–1
steady-state theory, 17
stereomicroscopy, 31
Stransky–Krastanov, 12
surface plasmon resonance, 140
surface quality
  and interface quality, 118–19
    in situ SE during CdZnTe substrate preparation, 119
    surface roughness and bulk thickness measured by in situ SE during microcrystalline silicon film growth, 120
surface X-ray diffraction
  epitaxial film growth, 239–68
  future trends, 267–8
  growth kinetics studies of pulsed laser deposition, 245–9
  SrTiO$_3$ PLD case study, 250–67
    average island spacing, 253
    crystallization timescale, 259–60
    diffuse scattering, 264–7
    growth stages, 252–4
    persistent growth intensity oscillations, 251
    quantitative determination of interlayer transport components, 260–4
    simultaneous two-layer growth, 254–9
    surface transport timescales, 251–2
SXRD transients, 250

thermal diffuse scattering, 31
thin film growth
  in situ characterization, X-ray photoelectron spectroscopy (XPS), 75–95
  buried interfaces using high kinetic energy XPS (HAXPES), 92–4
  in situ monitoring, 83–7
  thin film reaction measurement with gases using ambient pressure, 88–92

in situ characterization using RHEED, 3–26
  and pulsed laser deposition (PLD), 3–5
  intensity variations and Monte Carlo simulations, 23–4
  kinetical growth modes and intensity response in RHEED, 18–22
  kinetics vs thermodynamics crystal growth, 12–13
  pattern analysis: surface disorder, 8–12
  specular intensity variation during deposition, 13–18
in situ characterization using ultraviolet photoemission spectroscopy, 55–73
  future trends, 72–3
  principles, 56–63
  UPS applications to thin film systems, 63–72
in situ monitoring, 83–7
  detailed Si 2p and W 4f spectra taken after 11 and 61 minutes, 87
  schematic view, electrostatic lens system, 85
  W 4f, O 1s and Si 2p spectra obtained in situ and integrated XPS peak intensities, 86
  XPS spectrometer set-up, thin film layer deposition, 84
in situ spectroscopic ellipsometry, 99–144
  characterization, 109–19
  ellipsometry principles, 100–9
  in situ considerations, 119–31
  in situ examples, 132–43
reaction measurement with gases using ambient pressure, 88–92
  ambient pressure XPS O 1s, Ag 3d, and C 1s spectra taken at relative humidity, 91
  APXPS instrument principles, 90
  illustration reaction of MgO(100)/Ag(100) film with water vapour, 93
  integrated O 1s peak intensities from ambient pressure XPS, 92

volatilization and phase transition reactions, 88
real-time *in situ* surface monitoring by spectroscopies combined with RHEED, 180–209
  comparison and future trends, 206–9
  comparison *in situ*, real-time spectroscopies using the RHEED electron beam as excitation, 208
  in situ spectroscopies, 191–206
  processes and excitations by primary electrons, 181–8
  recombination and emission processes, 188–91
specular intensity variation during deposition, 13–18
  intensity oscillations during homoepitaxial growth, 14
  origin, 13–16
  other growth-induced variation, 16–18
  specular RHEED intensity modulation and aliasing effect, 17
thin multicomponent films
  in situ ion beam surface characterization, 155–76
  experimental set-ups, 161–5
  ion backscattering spectrometry and TOF ion scattering and recoil methods, 157–61
  studies of film growth processes, 165–75
three-step model, 60–1
time-of-flight ion scattering, 157–61
'tooling factor,' 214
toroidal ion energy analyzer, 162–3
  illustration, 162–3
total internal reflection ellipsometry, 140
total reflection angle X-ray spectroscopy
  increasing surface sensitivity, 195–7
  beam geometry, 196
  critical angle measurement, 196–7
  multilayer thickness measurement, 197
transmission electron microscopy, 31, 118
transmission Kikuchi patterns, 31
trimethylaluminum, 171

tunable diode laser, 224
two-event model, 30

ultra-high vacuum, 64, 109
ultraviolet photoemission spectroscopy
  applications to thin film systems, 63–72
    ARPES measurements, near E_F spectra $La_{1.6}Sr_{0.4}MnO_3$ thin films, 71
    c-axis lattice constant and resistivity and residual resistivity ratio, 65
    in situ ARPES measurements, $La_{1.6}Sr_{0.4}MnO_3$ thin films valence band, 70
    in situ photoemission valence spectra of $SrRuO_3$, 68
    $La_{1-x}Sr_xMnO_3$, 69–72
    near E_F spectral weight dependence on Ru stoichiometry, 66
    $SrRuO_3$, 63–9
    $SrRuO_3$ films resistivity, resistivity dependence and ferromagnetic transition on monolayer film thickness, 67
  principles, 56–63
    ARPES, 57–61
    high-resolution photoemission spectroscopy, instrumentation and light sources, 62–3
    Scienta analyser on beamline, 62
  thin film growth, *in situ* characterization, 55–73
  future trends, 72–3
  ultraviolet photoemission spectroscopy (UPS) principles, 56–63
  UPS applications to thin film systems, 63–72

vacuum ultraviolet, 61, 129
vapour flux monitoring, 212–36
  case studies, 231–5
    block diagram of atomic absorption system, 235
    chamber pressure dependence of sensors, 234
    co-deposition of Y, Ba, and Cu, 232–4
    flux monitoring of atomic oxygen and nitrogen, 234–5

MBE growth of oxide films with AA rate monitoring, 231–2
  pressure dependence of the control QCM, 233
optical absorption spectroscopy techniques, 223–9
  hollow cathode lamp vs tunable-diode laser source, 224
  overview, 212–14
    schematic of PVD vacuum system, 213
  quartz crystal microbalance, 214–17
  techniques and resources, 230–1
    comparison of deposition vapour sensors, 230
  vapour ionization techniques, 217–23
vapour ionization techniques, 217–23
  electron impact emission spectrometry, 222–3
  generic vapour ionization sensor, 218
  ion gauge and chopped ion gauge, 217–20
    schematic of ion gauges, 218
  quadrupole mass spectroscopy, 220–2
  schematic of a generic vapour ionization sensor, 218
virtual interface, 111–13
  virtual interface approach can simplify *in situ* SE data analysis, 113
volatilization, 88
Volmer–Weber, 12

wavelength dispersive spectrometer, 193
wavelength-modulation spectroscopy, 228

X-ray diffraction, 5, 245–6
X-ray emission spectroscopy, 189
  combined with RHEED, 193–5
  experimental set-up, 193–4
  quantification of X-ray line spectra, 194–5
X-ray photoelectron spectroscopy, 57
  in situ characterization of thin film growth, 75–95
  in situ monitoring, 83–7
  inelastic mean free path, kinetic energy function of electrons in copper, 76
  measurements of buried interfaces using high kinetic energy XPS (HAXPES), 92–4
  multilayer system model, homogeneous layers above bulk substrate, 79
  schematic depiction of angle-resolved XPS experiment for chemical states measurement and composition of SiO_xn_y films and Si 2p XPS spectra, 81
  Si 2p spectra at different photoelectron kinetic energies and transmission electron microscopy cross-sectional images of film, 82
  Si 2p XPS spectrum for SiO_2/Si and SiO_2/Si interface model, 78
  thin film reaction measurement with gases using ambient pressure, 88–92
  XPS experiment set-up, 76
X-ray photoemission spectroscopy, 171
X-ray photons, 187–8
X-ray scattering theory, 246

ZAF procedure, 195

CPSIA information can be obtained at www.ICGtesting.com
Printed in the USA
LVOW031019221111

256076LV00002B/5/P